Reading Iraqi Women's Novels in English Translation

By exploring how translation has shaped the literary contexts of six Iraqi woman writers, this book offers new insights into their translation pathways as part of their stories' politics of meaning-making.

The writers in focus are Samira Al-Mana, Daizy Al-Amir, Inaam Kachachi, Betool Khedairi, Alia Mamdouh and Hadiya Hussein, whose novels include themes of exile, war, occupation, class, rurality and story-telling as cultural survival. Using perspectives of feminist translation to examine how Iraqi women's story-making has been mediated in English translation across differing times and locations, this book is the first to explore how Iraqi women's literature calls for new theoretical engagements and why this literature often interrogates and diversifies many literary theories' geopolitical scope.

This book will be of great interest for researchers in Arabic literature, women's literature, translation studies and women and gender studies.

Ruth Abou Rached is an Honorary Research Associate at the University of Southampton and specialises in teaching Translation Studies, Modern Foreign Languages and Arabic studies to widen university access to under-represented groups. Her work on Iraqi women's literature was inspired by her community work in the UK and living in the Middle East. Her research interests include Iraqi and Arab women's writing, Palestinian and other exilic literatures, postcolonial studies and intersectional feminist translation theories. She is editor for *New Voices in Translation Studies*, International Association of Translation and Intercultural Studies (IATIS).

Focus on Global Gender and Sexuality

For a full list of titles in this series, please visit www.routledge.com/Focus-on-Global-Gender-and-Sexuality/book-series/FGGS

Reading Iraqi Women's Novels in English Translation

Iraqi Women's Stories

Ruth Abou Rached

LONDON AND NEW YORK

First published 2021
by Routledge
2 Park Square, Milton Park, Abingdon, Oxon OX14 4RN

and by Routledge
52 Vanderbilt Avenue, New York, NY 10017

Routledge is an imprint of the Taylor & Francis Group, an informa business

British Library Cataloguing-in-Publication Data
A catalogue record for this book is available from the British Library

Library of Congress Cataloging-in-Publication Data
A catalog record for this book has been requested

ISBN: 978-0-367-85717-2 (hbk)
ISBN: 978-0-367-56447-6 (pbk)
ISBN: 978-1-003-01456-0 (ebk)

Typeset in Times New Roman
by Apex CoVantage, LLC

Contents

Note on transliteration

For the transcription of Arabic language citations, this study follows the style used by ALA-LC (American Library Association – Library of Congress).

<https://ijmes.chass.ncsu.edu/docs/TransChart.pdf>

Titles of Arabic publications are listed in Arabic, with no ALA-LC transliterations. Titles are back-translated into English between square brackets for ease of reference.

For Arab authors with publications in a language other than Arabic, their names are kept in the form used with their publications. Arabic words or titles taken from authors' quotations are kept in the form transcribed by them.

Frontispiece

Mediterranean Summer (1989), Sarah Niazi

Girl Under the Moon (1988), Sarah Niazi

Girl Under the Moon (1988), Sarah Niazi

1 Pathways of Iraqi women's story-writing in English translation

At a literary event on Iraqi literature held in Berlin in 2020, Syrian-Palestinian poet Ghayath Almadhoun remarked that everyone in the Middle East, particularly Iraq, seems to be 'waiting in line' to tell their stories of war, dictatorship, survival and exile. Iraqi women writers have not been spending time waiting: for decades they have been writing their – other Iraqis' – stories of hope and survival, many engaging with translation as part of their creative expression and literary activism. Little, however, has been written on how stories written by Iraqi women writers have been mediated – or 're-written' – in English translation. This gap is surprising in view of Iraq's high international profile for decades. As a response to this gap, this book explores how six examples of Iraqi women's story-making have been mediated in English translation across different times, political contexts and locations. The writers in focus are Samira Al-Mana, Daizy Al-Amir, Inaam Kachachi, Betool Khedairi, Alia Mamdouh and Hadiya Hussein, six Iraqi women writers who have turned to translation as one way of preserving their stories – and stories of many other Iraqis – beyond conventional borders.

Samira Al-Mana and Daizy Al-Amir were amongst the first Iraqi women writers to publish stories in Beirut during the 1960s. Inaam Kachachi writes stories to archive alternative histories of Iraq as 'amulets of memory' (Snaije 2014). Betool Khedairi works to archive 'photos' of Iraq through her creative writing. For decades Alia Mamdouh has challenged the many languages of patriarchy via her innovative use of Arabic. Hadiya Hussein writes to show Iraq's beauty while addressing fear, specifically "the fears of an Iraqi citizen who has become even more fearful of a slip of the tongue than an actual, physical slip" (Hussein cf. Lynx-Qualey 2013). Drawing on analytical frameworks of intersectional feminist translation, I show how one story of each writer has moved across its many charged borders, some of which included changes of government, deteriorating infrastructure and ongoing conflict in Iraq. English translation is, of course, not the only way by which Iraqi women writers communicate their stories with different readerships, and story-writing is not the only way that Iraqi women express their literary activism. Innovative literature by Iraqi women, particularly during the early years of their story-writing yet to be translated or to circulate outside Iraq (Abou Rached 2020, 51–51) also merits much more critical attention. Poetry, theatre, art and music are also just four of the many rich categories of Iraqi expression which are beyond the scope of this book.

As part of a Series which seeks to provide a focused overview on topics to encourage further research, this book does seek to cover all stories by Iraqi women writers in translation in English and other languages. Underpinning the focus of this book on Iraqi women's story-writing through case studies in a specific context, however, is a consciousness that many innovative works are only accessible in specific locations or are no longer available for many reasons, amongst them censorship, war and varying locations languages and politics of publication. This book aims to raise wider appreciation of the critical importance of specific works though the nexus of translation to ensure the vital issues their writers raise across borders remain as a source of debate, discussion and inspiration. By exploring pathways of English translation for these six stories, this book invites wider research on Iraqi women's story-making as a body of creative activist work crossing many genres and borders.

The politics of reading Iraqi women's stories

While Iraqi women have always been a bedrock of Iraqi cultural memory, women writers have always been working within many "political and social restrictions to make their voices heard throughout the Arabic-speaking world" (Abu-Haidar 2005, 193). Successive wars in Iraq have shaped their cultural production, specifically the 2003 war and ongoing military occupation arguably 'marking' Iraq as the first 'new' neo/colonial Arab nation state of the twenty-first century (Ismael 2014). This war also presented Iraq as site of where a long-standing dictatorship could fall. Saadi Simawe (2004) has framed the ramifications of Iraq's history of dictatorship and war as oddly conducive to Iraqi creativity:

> It seems to me that literature and the arts live on disasters which, because they challenge our very existence, force us to think imaginatively and creatively. The African American experience produced the blues, the Jewish experience produced Kafka, and the Russian sorrows gave us Dostoevsky. Iraqis' tragedies have produced profound literature that needs to be published. . . . [T]here are now two Iraqi literatures: the literature of exile and the literature under fascism and war.
>
> (Simawe 2004)

He also saluted Iraqi women writers for "an existentialism ... mixed with humour alongside an unusual compassion for all humans," that is, their representations of everyday people in Iraq making choices in almost impossible circumstances. This way of writing is not new in Iraqi women's literature: Iraqi women have always written of how everyday Iraqis face interlocking effects of poverty, corruption, violence and social injustice. Shifting since 2003, however, is how some women writers view the act of writing. Haifa Zangana first described the post-2003 space as akin to a "sterile, dark silence extend(ing) its shadow over the imaginary," leaving many Iraqi writers, herself included, feeling that "the cruel reality of occupation has turned writing fiction into a meaningless act" (Zangana 2004, xv). Inaam Kachachi similarly distances violence as a source of her creativity in any language:

> Violence tires me, and my description of its mechanisms cannot match the skill with which it is being committed on the ground. Give me peace and I will describe it to you and furnish it with characters. Don't you think that a warm, well-lit room conveys the darkness to us?
>
> (Kachachi cf. Najjar 2014)

Sinan Antoon also configures translation as an act of mourning, insofar that the demand of Iraqi literature in English translation seems at times predicated on war and violence in Iraq rather than appreciation of the aesthetics of Iraqi creativity (Mattar 2019, 73). This 'demand' for violence – in translation – opens up the risk of a work going far beyond the intention of any Iraqi writer by whichever mode s/he uses to mediate her/his work. Iraqi artist Wafaa Bilal (Bilal & Lydersen 2013) discovered that spectators based in the United States were all too happy to fire paintballs at him at the click of a button during his interactive installation "Kill an Iraqi/Domestic Tension." Although his installation purposefully invited audiences to confront violence, the sheer force of the 'hits' he received raises many questions on what it means to constructively engage art and violence in Iraq with international audiences. Other questions relate to the impact of gender on Iraqi aesthetic and literary production inside and outside Iraq. What if Bilal had been a woman? Or with a group of women? Such questions relate to crucial debates within Iraqi scholarship, particularly in the wake of the US media – post-George W. Bush's Greater Middle East Initiative – saturating its publics with images of Iraqi women voting in post-2003 Iraq elections (Denike 2008; Winegar 2005) as Iraqi citizens "helping [to give] birth to freedom" (Wolfowitz 2004, cited in Al-Ali & Pratt 2009, 83), the midwife of this freedom being US army intervention and occupation. While not relating to translation, such debates suggest that an Iraqi women's story published in English translation is a charged and productive site of inquiry 'requiring' more in-depth critical attention and exploration.

One way of thinking through how Iraqi women's stories move across translation is to read the English versions as more than just a new 'version' of a 'first' Arabic-language story published in the wake of specific events in Iraq. Clearly Iraqi women write in response to particular political-historical junctures (Abdel Nasser 2018; Abdullah 2018; Atia 2019; Mehta 2006), as do Iraqi men, with more recent novels focusing on post-2014 presence of Islamic State of Iraq and the Levant (ISIL) in Iraq (Hamedani 2017). Majeda Hatto (2013) contends that Iraqi women's writing is far more than the poetics of women's experience being "بقايا ضلع رضوض من اضلاع الخطاب الذكوري" [remains of a bruised rib within the rib-cage of masculinist discourse] (2013, 3) on war and nationalism. In other words, the varying poetics of Iraqi women's writing have always expressed how power injustices relating to women could apply to anyone in Iraq (Abou Rached 2020). Lutfiya Al-Dulaimi's short story collection البشارى [*Glad Tidings*] (1974) for example showed dystopian worlds dominated by psychic uncertainty. In one story, women find themselves acting in a play they thought they were going to watch. Another woman is overjoyed to hear "البشارى" [the glad tidings] or 'good news' that she has run a red traffic light (1974, 43). Whether she is happy to be seen crossing a red line or to have identified what the red line/light is in the first place is left open to interpretation. Interweaving Bouthayna Al-Nasiri's short story "القارب" [The Boat] (from her collection حدوة حصان [Horse-Shoe] (1974)) about long-standing traditions in local river communities) is also a silent threat of violence pervading the world of each protagonist, woman or man. The impact of censorship is often alluded to in Iraqi women's stories. In ملتقى النهران [Where the Two Rivers Meet] (1968), Daizy Al-Amir writes of two women expressing solidarity through holding hands, neither woman having the power of speech. Maysalun Hadi's story "زينب على ارض الواقع" from the 1994 collection رجل خلف الباب [*A Man Behind the Door*] translated into English as / translated into English by Shakir Mustafa under the title of [Her Realm of the Real] (2008) presents a mouse asking about his missing beloved without stating exactly why a mouse is asking about a missing person in Iraq. Neither writer however states why their protagonists cannot speak openly. Other earlier stories by Ibtisam Abdullah, Salima Salih and May Muzaffar represent 'everyday' tragedies of Iraq in 'documentary' and poetic ways, their very occurrence obviating the need for overt political critique. Recent stories such as Hadiya Hussein's ما بعد الحب [*Beyond Love*] (2004) refer to specific tragedies of Iraq within gendered frames to ask more direct and pointed questions of their perpetrators.

Reading earlier examples of Iraqi women's writing in English translation can present challenges as the politics of many writers do not seem to be overtly expressed or articulated in the text. Stories by Iraqi women were firstly published in Iraqi literary journals such as *Iraq*, *Iraq Today* and *Gilgamesh* (Altoma 2010) sponsored by the cultural arms of a prosperous Iraqi government seeking to promote itself as a leader of pan-Arab power and solidarity. Publishing in English in such journals did not mean that an individual writer could express political opinions freely. With Iraqi President Saddam Hussein insisting "The *pen* and the barrel of a gun have one and the same opening" (Antoon 2010, 29), all Iraqis were subjected to intense state surveillance when the 1968–2003 Iraqi Ba'athist government ruled over Iraq, in whichever language they published. Those who did not confirm to state directives often simply 'disappeared.' Raad Mushatat (1986) recalls how poet Burhan Asshawi and story-writer Jihad Majid were imprisoned and tortured for their refusal to conform. Nejem Wali (2007, 52) recalls: "even those who chose to quit writing saw themselves forced to write something which did not rile the dictator, because even silence was considered a crime," meaning, in the words of Alia Mamdouh, that "the ban on speaking and the obligation to speak" (Mamdouh cf. Chollet 2002) was a site of productive and tragic tension for many Iraqi writers. Such issues of productive tension of silence and 'speaking' raise important questions on how we could approach reading Iraqi women's stories now: how to read such stories in English translation if we know little of their politics of writing, translation and production at the time?

While 2003 is not a 'watershed' for Iraqi women's writing *per se* it is fair to say that this date has been 'a line in the sand' for Iraqi women's literature in English translation published in the United States, at least. The prominence accorded to 'expert introductions' in many post-2003 US-based publications suggests that complex geopolitical, gendered power relations are (rightly) assumed to be interweaving new interest in Iraqi woman's story-writing in English translation. All four novels by Iraqi women writers published by the New York Feminist Press have critical introductions and afterwords by academic experts – Hélène Cixous and Farida Abu-Haidar in two novels by Alia Mamdouh (2005, 2008); Hamid Dabishi and Ferial Ghazoul for Haifa Zangana's *Dreaming of Baghdad* (2009); and Nadje Al-Ali for Iqbal Al-Qazwini's novel *Zubaida's Window* (2008). Hadiya Hussein's *Beyond Love* (2012), by Syracuse Press, has two introductory chapters by miriam cooke and translator Ikram Masmoudi. Iraqi women's novels shortlisted for the International Prize for Arabic Fiction (IPAF) in English translation, however, have no introductions at all. Inaam Kachachi's story about an Iraqi woman interpreter in the US army, *The American Granddaughter* (2011), has a note on its IPAF short-list status and a brief translator bio. The 2018 IPAF winner *The Baghdad Clock* (2018), by Shahad Al-Rawi, highlights solely its IPAF status and briefly states that it "takes readers beyond the familiar images in the news." Reading these two 'award' stories in English translation alongside the other novels framed with so many para/textual supplementations raise further questions: what local and global politics could be at play in the mediation of Iraqi and Arab women's story-making as 'glocal' – that is local and global – publications?

Such potentially varying politics of reception bring me to why I draw on intersectional perspectives of feminist translation studies to read six Iraqi women's stories moving across Arabic and English. It is a pertinent question to ask in view of the charged discourses of 'feminism' at play in post-2003 Iraq. Haifa Zangana reminds us that US state-funded 'feminist' NGOs in Iraq, often serving US state interests rather than the needs of Iraqi women, have significantly damaged the legacy of Iraqi women's local gender-focused political activism (Zangana 2005; Ismael 2014). The assassination of Iraqi woman poet and journalist Behjat Atwar in 2006 by unnamed 'opposition' militias (Zangana 2014) is a tragic example of how power relations in Iraq go far beyond definitions of what 'feminist' agency is or isn't in any language. One starting point is to consider how analytical frameworks of feminist translation could relate to Iraqi women's story-making. As an activist praxis, feminist translation interrogates 'the feminine' in a variety of ways: as a gender construct, women's experiences, linguistic relations of power and a metaphor for translation itself (Abou Rached 2017, 199). In earlier feminist translation scholarship, praxes of 'translation' were often explored – to expose and question – the patriarchal premises on which many languages and terms of reference are based, translation one way of rewriting them to transform them (Massardier-Kenney 1997; Simon 1996; Flotow 1991; De Lotbinière-Harwood 1991; Godard 1989). In more recent feminist translation studies scholarship, the interlocking relations of power such as those of race, ethnicity and class (Crenshaw 1991) have emerged as vital lines of inquiry alongside gender and other sites of discrimination, such as "racism, capitalism, colonialism, heteronormativity, ableism and so forth" (Castro & Ergun 2018, 125). Concerning translation, sites of discrimination and marginalisation could include language, location, cultural capital and sociopolitical identity. This is why critically engaging feminist translation with translated literary works does not mean that texts or writers of inquiry are identified as having a defined feminist or activist ideology. One fundamental premise of feminist translation analysis is actually to challenge categorical definitions of 'feminist' in the first place (Flotow & Kamal 2020; Castro & Ergun 2017; Flotow & Farahzad 2016; Álvarez 2014). As many Iraqi women's stories call many theories and critical methodologies into creative question, I engage with analytical perspectives of feminist translation to explore, for the first time, critical aspects of Iraqi women's stories in translation while considering such theories' geopolitical scope.

Any analytical engagement with feminist translation nonetheless calls for some clarification on how 'feminist' or النسوية [*al-nisūwiyya*] writing can be understood. In Arabic, النسوية [*al-nisūwiyya*] means "feminine," "womanly" (Wehr 1994, 1130) as well as 'woman-focused,' 'by a woman' or 'feminist.' Its potential slippages of meaning give some critical perspective to the negative connotations of النسوية [*al-nisūwiyya* in post-2003 Iraq] and how women's political agency is often configured within politically gendered frames of reference. It also contextualises why the term النسوية [*al-nisūwiyya*] is used by Iraqi scholars writing in Arabic about Iraqi women's literature (Khodeir 2013; Hatto 2013; Kadhim 2017; Al-Dulaimi 2016; Ahmad 2017). Hadil Ahmad uses النسوية [*al-nisūwiyya*] as a strategic frame for defining Iraqi, Arab and women's writing as a specifically gendered frame of interaction:

"المعلوم ان ما تكتبُ المرأة من الأدب كان وما زال مثار الجدل واسع والإشكالية لا تنتهي.
. . . تبدأ التسمية (الأدب النسوي) وتمر بخصوصية الجنس الأنثوي المقابل الهيمنة الذكورية والتنميط الثقافي للمرأة في المجتمعات العربية، ولاتنتهي بالموضوعات التي تتضمنها النصوص الإبداعية، ونعني بها المضامين التي تدعو الى التمرد على التقاليد السائدة الموروثة والتحرر من السلطة الأبوية."

[Whatever literature a woman writes has evoked – and continues to evoke – a sense of controversy and ambiguity which never ends. . . . It starts with literature being labelled as '*nisūwī*', and how it relates to the specificities of feminine gendered experience in the face of masculinist hegemony and the cultural stereotypes about women in Arab societies. It does not stop at themes covered in creative literary texts, that is, the very themes which call for rebellion against the prevailing customs passed down through the generations and call for liberation from patriarchal authority.]

(Ahmad 2017, 7)

While Ahmad does not refer to the languages by which Iraqi women's writing is mediated, her interpretation of النسوية [*al-nisūwiyya*] writings evokes questions on what it means to *read* Iraqi women's stories in English translation as part of wider chains of discourses connecting to the politics of patriarchy and solidarity in other languages. Questions such as: for whom women's writing evokes such controversy and ambiguity and who is doing the labelling? How would such questions relating to النسوية [*al-nisūwiyya*] story-making in Arabic 'move' across languages?

Without claiming to have any definitive answers to such a wide-ranging question, I seek to engage with and further conversations about Iraqi women's stories by focusing on six different novels published across diverse times and contexts in English translation. The first chapter focuses on 'the politics of the uncanny' in Iraqi literature and explores حبل السرة [*The Umbilical Cord*] (1990) by Samira Al-Mana and Daizy Al-Amir's على لائحة الانتظار [*On the Waiting List*] (1988) as two early Iraqi women's stories in English translation. Moving to post-2003 Iraq, I consider the mediation of الحفيدة الأميركية [*The American Granddaughter*] (2009) by Inaam Kachachi in English translation from analytical frameworks of 'audible' feminist translation praxis. I then read Betool Khedairi's polyphonic novel كم بدت! السماء قريبة [*A Sky So Close!*] (1999, 2001), which mixes Iraqi dialect, formal Arabic and English as a conversation on the politics of Iraqi marginality in the 1970s across the pre- and post-9/11 temporal contexts. I then explore how the English version of حبات النفتالين [*Mothballs*] by Alia Mamdouh (1986/2000) and ما بعد الحب [*Beyond Love*] (2003) by Hadiya Hussein confront translation in ways which invite us to consider and question what it means to read Iraqi women's stories of conflict across languages in productive and transformative ways. There are no definitive 'theories' arising in this book: many of the questions co-emerging in my analysis are testament to the depth and complexities of Iraqi women's stories. Their pathways of translation from Arabic into English are just one of many vital critical points of focus. My hope is that this brief study is one of many further engagements with the diverse aesthetics of Iraqi and women's literature in contexts of translation and other formats of literary expression.

2 Translating 'the uncanny'

Samira Al-Mana and Daizy Al-Amir

Both born in 1935, Samira Al-Mana and Daizy Al-Amir are amongst the 'first generation' of Iraqi women story-writers. They grew up in Iraq when the whole country was going through huge political changes independence from British colonial influence; the rise of leftist politics in Iraq in the 1950s; increasing influence of nationalist discourses leading to the Iraqi Ba'athist party assuming political control of Iraq in 1968, which would last until 2003. Both writers published their first stories during the 1960s, a time when the Iraqi Ba'athist Party was first rising to power and Iraq was becoming known as a centre of cultural influence. While Al-Mana and Al-Amir lived many years outside Iraq, both writers dedicated their story-writing to recording every-day Iraqi experience. They did this during times when most writers could not explicitly express their individual political opinion in the Iraqi public sphere. Shakir Mustafa, for example, likens Samira Al-Mana to a "modern-day Scheherazade intrigued by story-telling but mindful of the risks it involves" (Mustafa 2004, 129). Daizy Al-Amir is acclaimed for expressing the voices and silences of women from various perspectives (Ghazoul 2008, 193), often through metaphors of travel, transit and (forced) departure.

Samira Al-Mana's حبل السرة [*The Umbilical Cord*] (1990) was the first novel to focus on the 1980–1988 Iran-Iraq war as it was experienced by diaspora Iraqis living in London. Daizy Al-Amir's 1988 short story collection على لائحة الانتظار [*On the Waiting List*] was the first to present experiences of war in Lebanon from the perspective of one Iraqi woman. The subtleties of the politics expressed in both works present a challenge to read in English translation as much of their politics is expressed in 'uncanny' or oblique ways. To give background to why this is the case, I situate how the leitmotif of 'the uncanny' in post/colonial literary writing relates to Iraqi story-writing and feminist translation. I then analyse how 'حبل السرة [*The Umbilical Cord*] (1990) by Samira Al-Mana and Daizy Al-Amir's على لائحة الانتظار [*On the Waiting List*] (1988) move across Arabic and English using perspectives of feminist translation which focus on reading 'uncanny' presences and absences in translation yet to be explored in contexts of Iraqi women's story-writing.

The 'uncanny' and الإغتراب [*al-ightirāb*] as home, exile and 'normality' in Iraqi literature

Sigmund Freud (1919, 240) described 'the unfamiliar' or 'uncanny' as "an emotional impulse" related to a fear or an anxiety repressed coming to light in the human psyche through a person seeing or experiencing something odd or 'off-tilt' as strangely familiar – but without knowing why. In the context of postcolonial literary studies, Homi Bhabha (1992) has drawn on Freud's notions of 'the uncanny' to explore how covert or 'liminal' representations of fear, anxiety and political resistance emerge in "the house of fiction" (1992, 141) as responses to wider asymmetries of political oppression underpinning the meaning-making of their works and their contexts of mediation (Bhabha 1994, 15). According to Bhabha, this means that a work can making c/overt references to – or liminally enacting – languages, literary traditions and histories erased, distorted or muted by colonisation or totalitarian rule while 'presenting' as a perfect example of literature within a certain canon of writing. Bhabha takes the example of a writer mimetically conforming to totalitarian or colonial directives while somehow not reading, in particular contexts, as 'quite right' – or in the postcolonial contexts of India that Bhabha was referring to, quite "white" (1994/2004, 128) – to illustrate his point. Writers can also evoke themes of resistance through fusing elements of the impossible and the everyday in their stories to allude to tensions of the 'abnormal' being 'the new normal.' They often blur the genres of 'fiction' and 'nonfiction, 'personal' and 'collective' alongside 'private' and 'public' in subtle ways. The existence of such tensions is not usually explicitly stated in such 'uncanny' modes of story-writing. Under situations of extreme censorship, a writer may conform to 'official' directives but work to invite alternative interpretations of her/his writing (1994/2004, 138) in ways which will not incriminate him/her or others. This is why 'the uncanny' in literature can operate as a powerful frame of resistance and a critical site of political action.

Arab writers have a long tradition of making recourse to allegory, parable, sci-fi and magic realism to represent the 'unfamiliar' as uncannily 'familiar' to their different readerships, censors included. Arguably, *1001 Nights* is one early example of 'science-fiction' in Arab world circulation whose appeal has lasted for centuries due to the multivalent ways in which each story can be read. Hassan Blasim (2016) describes the prevalence of 'uncanny' writing in contemporary Iraqi literature in particular as much more than allegory: it is testimony to Iraqis experiencing a "long saga of wars, death, destruction, population displacement, imprisonment, torture, ruin and tragedies" (2016, v). Strange events or events of horror in stories thus often do not read as 'strange' to most (Iraqi) readers but are recognised as, tragically, already experienced by them or others. During earlier times of very directive censorship in Iraq, allegory was also "a highly pointed, hard to misappropriate and aesthetically appealing" (Simawe 2009, 129) tool of survival for Iraqi writers. The leit-motif of الإغتراب [*al-ightirāb*], for example, was one prevalent theme in such examples of Iraqi story-writing. In Arabic, the word الإغتراب [*al-ightirāb*] can mean feeling separated from one's homeland and can also mean *feeling* strange or alienated within one's society, life and oneself. The literal meaning of الإغتراب [*al-ightirāb*] allowed an Iraqi writer to overtly express patriotic identification with the (Iraqi) 'homeland' while c/overtly express-ing the *psychic* frames of feeling 'strange' and alienated by everyday reali-ties in Iraq. This and other presentations of 'the uncanny' in earlier Iraqi story-writing (Ghazoul 2004) related realities of publishing inside Iraq too.

Censorship – and fear of it – was certainly one barrier to critical recep-tions of earlier Iraqi women's story-making in Iraq (Al-Mana & Abou Rached 2017; Abou Rached 2020). Opportunities for Iraqis to show their work to others outside Iraq were also very few and far between during the times when many Iraqis were not permitted to travel outside of Iraq during the 1980–1988 war (Al-Amir 1994) and after international sanctions were imposed on Iraq from 1991 (cooke 1997). This point gives vital context to why Texas University Press editor Annes McCann-Baker explained that Daizy Al-Amir's stories would be translated "as quickly as possible" after meeting with her just once in Texas in 1989 (McCann-Baker 1994, vii). In later years, more stories by Iraqi women writers were published in Eng-lish translation within Anglophone academy contexts, often accompanied with short, carefully worded translator introductions giving some – but not detailed – critical context. In her introduction to her translation of Aliya Tayib's stories, miriam cooke (Talib 1994, 80) seems reticent to write much about the author at all. She simply explains that she knows very little about Tayib and urges the reader to read her stories in English "carefully" without giving further guidance on what 'reading carefully' means. Similarly, in *Final Night* (2001), a collection of Bouthayna Al-Nasiri's stories which had been published in Arabic over different decades, translator Denys Johnson-Davies (2001, 1) compares Al-Nasiri's writing style as "more direct and relaxed" at times but more tense in others. He also does not expand further about why this was the case (Abou Rached 2020, 52). We could read both translators' silence as working to protect an Iraqi writer from any politi-cal repercussions arising from her work appearing in English translation. After all, English was one of the many hegemonic languages of power used to mediate discourses of war in Iraq from 1990. Translator silences could be read in other ways: a translator or editor just wanting to present a new genre of Iraqi women's literature to new audiences with as little 'clutter' as possible. These two examples highlight the challenges of reading c/overt meanings of stories where nothing 'concrete' is explicitly explained. How to read the politics of silence in Iraqi women's stories if nothing definitive is stated? How much of a story's import and meaning can be gleaned in any language without a reader having extensive (political) knowledge of the time?

Mediating silences, politically motivated or otherwise, is not a new phenomenon for many translators of literatures of women and other marginalised groups. Feminist translators have historically engaged with *gendered* ambivalence and political marginality in different ways. One way is through individual translators, editors or publishers opting to translate *particular* writers' works or particular canon or languages of literature as a way of widening and shaping the canon in which authors and works can be found (Massardier-Kenney 1997, 59–60; Shread 2011, 283). The Arab Women's Writers' Series (1995–1998) edited by Fadia Faqir, published in the United Kingdom, is one example of this practice. As a transnational collaborative project, this series presented acclaimed novels by Arab women writers and autobiographical essays by many writers from MENA (Middle East and North Africa) in English for the first time. Another example of 'canon'-focused activism is how Iraqi women's literature in English translation first appeared largely through the efforts of scholars and editors such as Shakir Mustafa, Samuel Shimon, Ferial Ghazoul, Farida Abu-Haidar and miriam cooke (Abou Rached 2020, 52). Other approaches involve a translator overtly showing her/his role as a co-producer of a work's new meaning *and* as a political agent pursuing particular agendas in the textual creation process, usually by prefaces, afterwords or footnotes (Massardier-Kenney 1997). A translator making overt what her/his agenda helps to ensure, in the words of Françoise Massardier-Kenney, that (feminist) translator interventions "mean more than moving into new territory and acting exactly as the male tradition has acted" (1997, 63) – that is, 'invisibly' appropriating and overwriting a work without showing the dynamics of power at play in this process (Castro 2009). One example of counter-patriarchal translator practice is Carol Maier openly explaining in her Afterword to *Memoirs of Leticia Valle* why she carried out particular interventions, fully conscious that she was translating into English a literary work about one woman seeking to subvert patriarchy first written in Spanish (1994, 188). Another example is the Translator's Note to Alia Mamdouh's novel *The Loved Ones* (2008), in which Marilyn Booth takes great care to explain how the novel's 'multilingual' and 'transnational' themes are reflected in its linguistic and cultural references while crediting the writer Alia Mamdouh herself for assisting with the translation of this "polyphonic" work (2008, 277).

Critical debates on translator 'visibility' provoke questions on why explicitness of expression seems to be privileged in feminist translation studies scholarship that engages with genres of re/writing often shaped by marginalisation. After all, many translated works are not accompanied by prefaces or afterwords to explain any 'gaps' in their translation. As Barbara Godard (2002, 65) points out, however such 'gaps' are often manifestations of 'the differend' – 'the differend' meaning what cannot be expressed due to prevailing discourses precluding the possibility of their expression in the first place. One gap would be why a translator presents a literary work as important to read in English while not explicitly stating her politics of translation. Godard explains how such gaps or 'differends' can be explored in new contexts as follows:

> La traductrice souligne la différence – le différend même – entre les contextes de "l'original" et de la traduction, délimite paramètres du processus de transfert et explique la modalité de circulation du texte traduit dans son nouvel environnement.

> [The woman translator underscores the difference – the differend even – between contexts of "the original" and the translation. She sets out the parameters of the transfer process and explains the translated text's mode of circulation in its new environment.]

> (Godard 2002, 65)

Here we see, Godard does not specify where and how a translator should 'underscore' the difference, or to whom s/he should 'explain' a text's mode of circulation in its new settings. She also does not state who 'the translator' is. Patricia Hills Collins (2017, xi) saliently points out that many expressions of (activist) translator or writer agency *cannot* always be 'readable' in the ways s/he intends due to dissonant epistemes of intelligibility, whichever the language of mediation. So, in one critical context, a translator's silence on her own politics could be understood as 'not speaking out' in the face of oppression. In another situation, such silence can be 'read' as taking an ethical stance to protect others within their different lines of political vulnerability. Such conceptualisations of translator intervention framing 'the differend' *as part of* a translated work help us diversify and interrogate how and where (potential) activist 'gaps' could be discerned as 'readable' in a translated work: via the genre or type of work that is being translated? In aesthetic formats? Which language is it being translated to and from? Such questions situate the surrounding materials presenting a translated work such as front covers, introductions, footnotes and blurbs, defined as 'paratexts' (Genette 1987; 1997) or 'paratranslation,' as crucial locations of translator political action. And while the politics of paratexts is not a new field of inquiry for feminist translation studies (Massardier-Kenney 1997; Wallmach 2006; Castro 2009) or paratranslation studies (Garrido 2005; Frías 2012), notions of 'the differend' as non-explicated 'gaps' as spaces of activism open questions on where the 'text' and 'paratext' should be understood to be in the first place.

Questions of where translator action can – and cannot – be read to be pose exciting and challenging questions when reading حبل السرة [*The Umbilical Cord*] (1990) by Samira Al-Mana and Daizy Al-Amir's على لائحة الانتظار

[*On the Waiting List*] (1988). While little about the translation method is mentioned in each work, small but significant para/textual traces such as epigraphs and editor notes intimate something of the activist processes involved. To explore the importance of such traces, I consider different aspects of each work from an analytical framework of 'feminist paratranslation' (Abou Rached 2017), which explores and interrogates definitive notions of text and paratext, translation and paratranslation, translator and paratranslation by blurring definitive borders between them, just as feminist translation blurs the borders between 'writing,' 're-writing' and 'translation.' Reading an activist work as a para/translation and para/text (rather than as translation or 'text') interrogates 'unitary' configurations of 'authorship' and 'translatorship' by configuring editors, publishers and readers alongside 'writers' and 'translators' as potential 'paratranslators' of a work's meaning-making (Abou Rached 2017, 200). Not explored in so far is how useful such paradigms of para/translation are for reading manifestations of para/translator agency c/overtly concealing disclosure of political intention in the context of earlier Iraqi women's literature. In this chapter I read 'liminal' or 'uncanny' action taking place in the para/texts of حبل السرة [*The Umbilical Cord*] (1990) by Samira Al-Mana and Daizy Al-Amir's على لائحة الانتظار [*On the Waiting List*] (1988) using analytical frameworks of feminist para/translation. I explore the politics of recovery moving across both versions of Al-Mana's حبل السرة [*The Umbilical Cord*] (1990); I then discuss how Al-Amir's aut/her/ship is presented in the para/text of *On the Waiting List: An Iraqi Woman's Tales of Alienation* (1994).

Samira Al-Mana and حبل السرة [*The Umbilical Cord*] (1990): recovering lost stories

Samira Al-Mana has lived more than 30 years outside of Iraq (Mustafa 2004, 129), mostly in London. She has written six novels, three collections of short stories and one play. Along iconic Iraqi writers Fu'ad Al-Tarkarli, Mahdi Isa Al-Saqr and Gha'ib Tu'ma Farman, Al-Mana has been acclaimed in Iraqi literary scenes for taking earlier Iraqi story-writing beyond "الأطار 'تقليدي' ["a traditional frame"] of writing (Thamer 2004, 7). In 1984, Al-Mana and her husband, Iraqi poet Salah Niazi, set up the Iraqi literary journal الإغتراب الأدبي [*Literature in Exile*] (1985–2002) to provide a literary platform for diaspora writers and literary translators, regardless of their political position, stated or unstated (Niazi & Abou Rached 2020). The esteemed status of Samira Al-Mana in the diaspora Iraqi community is reflected by the Lifetime Achievement award presented by The Humanitarian Dialogue Foundation in May 2018. The novel حبل السرة [*The Umbilical Cord*] (1990) reflects, in many ways, aspects of Al-Mana's own life living as an exile from

Iraq connected to many literary circles outside of Iraq. This story is about a community of Iraqis living in London during the time of the Iran-Iraq war (1980–1988). The stories in this story are told from the perspectives of two Iraqi women, Madeha and Afaf. Its title حبل السرة refers to the attachment that Iraqis feel towards their country and peoples, despite the war, surveillance and Iraqi state directives often creating fear and distrust between them. Afaf remembers her relatives in Iraq warning her: "You must only look, only look without comment" (2005, 30) when she visits Iraq. In London, Madeha observes a politics of 'looking without commenting' amongst Iraqis living in London too, meaning that many Iraqis feel الإغتراب [*al-ightirāb*] wherever they are and with whomever they meet. In this novel, Al-Mana shows the gradual process by which Iraqis living in London during the 1980s may have started to look and comment on their sense of الإغتراب [*al-ightirāb*] and seek to recover from it. She does this by showing how Afaf and Madeha listen to and tell stories as a way of confronting their own fears and the fears of others.

From an analytical perspective of feminist paratranslation, I begin my reading of the novel's outer front cover or jacket as the first page and key site of analysis. The image of a latticed window is used on both versions. The 1990 cover depicts opaque blackness behind the window; in the 2005 English version, a mixture of warm colours. While the difference could be due to practical formatting reasons, the impact is striking: the black and white version (1990) intimates a bleak space; the English, a potential warmer home-space. The reader-window relationality is ambivalent: the reader could be outside looking in or vice versa. The jacket of the Arabic version gives no information on the novel, apart from a brief authorial biography of Samira Al-Mana. The back jacket of the English version introduces the two main Iraqi women protagonists Afaf and Madeha as two women "uprooted from the home because of the troubled events in the recent history of Iraq." It also explains the story is set during the 1980–1988 Iran-Iraq war and how "Afaf and Madeha learn through the British media how thousands of their people are being slaughtered and their country destroyed." This explanation is a reference to the geographical location of where both women receive news of Iraq – London – in a pre-digital media era and how many Iraqis could only found out about the extent of the war's devastation through alternative sources, as the Iraqi state media only represented it through 'official' reports of heroic victories enacted by the military (Khoury 2013; Rohde 2010). Little is stated about the novel's translation, other than it was translated by Samira Al-Mana herself, with help of the editor Charles N. Lewis. Samira Al-Mana in fact self-translated her work by reading it out loud in English, and Charles Lewis transcribed and edited what she said without knowing any Arabic (Al-Mana & Abou Rached 2017). A tragic para/text to its English para/translation is that Lewis died suddenly before its publication, its trace openly noticeable by the capitalised epigraph – **IN MEMORY OF THE LATE CHARLES N LEWIS** – in the first inner page. Other traces of this loss are manifest through minor typos (Lewis's pre-proof edits). Out of respect for Lewis, Al-Mana did not change his work or ask anyone else to carry out final edits or proof-reading (Al-Mana & Abou Rached 2017).

The story first opens with Madeha, the wife of a diplomat who lives in London. She translates for Iraqi publishing houses, attends Arab women's cultural events and holds soirées for Iraqis in London. She refrains from making open political commentary. The reasons for this are made clear where we read the chapter where Madeha accompanies an Iraqi man (for whom she has just interpreted) on his visit to a colleague before the Iraqi man's return to Iraq. After it dawns on Madeha that the man is visiting his colleague only to spy on him, her shock presents a sense of why some Iraqis in the 1980s never felt safe even in their own homes. For this reason, Madeha also suspects a trusted family friend to be an informant. When this friend later confronts Madeha about her suspicious attitude towards him, she justifies her fear:

"صدقني اننا صدى لما نسمع ونرى، لقد كاشفتني بصراحتك المعهودة في مكان وزمان يتردد المرء فيه من الحديث السياسي حتى لأقرب الاصدقاء اليه. كنت قلقة وخائفة على اولادي اكثر مما كنت قلقة على نفسي، انت تعرف ذلك ولست جاهلاًبالاوضاع العامة"
(1990, 60–61)

"Believe me, we're all echoes of what we hear and see. And you opened your heart to me in your candid way in a time and place where one would be hesitant to enter into political discussion even with one's nearest and dearest. I was frightened for my sons more than for myself. You know what I mean, you're not unaware of the general situation."
(2005, 60)

As Madeha gives no explanation of what the 'general situation' refers to, a reader is left to work out (or not know) her/himself what it may be. In the Arabic version, the next paragraph follows on directly from this excerpt. In the English version, there is a distinct break, as if the chapter has ended. From a perspective of feminist para/translation which explores and interrogates notions of para/texts as 'outside' of a work's meaning-making, this paragraph break could be significant in light of 'the current situation' of Iraqi writers publishing inside and outside of Iraq at that time feeling called to 'speak' and 'not speak'(Mamdouh c.f. Chollet 2002). This paragraph break could be read as a silent para/translatory intervention to invite reflection on what 'the general situation' in Iraq is/was. Or it could be read as an unintentional typesetting. Reading this paragraph 'break' as a para/text and para/translation draws attention to the importance of questioning the 'construction' of the story as a self-para/translated, self-published work: was the break a conscious choice by Al-Mana? A typesetting by Charles Lewis? A decision of the printer? While the gap does not cause a dramatic shift in the novel's meaning-making, the presence of this break highlights the importance of reading what appear to be innocuous para/textual mediations in this story as an echo of what we may miss in this novel's complex web of co-collaborative meaning-making as it moves from Arabic into English.

One of the most important 'silent' echoes in this novel relate to the un/ heard voices of Iraq's past: Iraqi women who never had the opportunity to learn to read and write their own stories. Traces of these stories are mediated via Afaf, who like Madeha, works in London. Much of Afaf's own story is linked to the (as yet) untold stories about the injustices suffered by her grandmother and her Iraqi foremothers, the first of which she hears from her elderly uncle telling stories of his youth in turn-of-the-century Iraq (1990, 42; 2005, 38). Afaf herself has been long divorced from her husband Jalal, a leftist political activist 'blind-sided' by the rise of the Iraqi Ba'athist party from the 1960s. By slowly connecting long traditions of patriarchy to her own trials with her husband, Afaf soon realises that the concerns and stories of individual women in the domestic sphere are not petty matters of 'less importance' than the national issues preoccupying her husband and many other Iraqi intellectuals. She works out that Iraqi women's stories are symptomatic of them. We read how lines of Afaf's memory of her own past stories begin to slowly impact on how she lives her present and so tries to reconfigure a different political future. The first 'line of memory' is when Afaf remembers nearly breaking her back when repairing a curtain that her husband Jalal would not repair, due to his concern with 'bigger' (leftist) political problems (1990, 81; 2005, 82). When she falls ill again, Afaf suddenly remembers what her uncle had told her about the injustices towards her foremothers in her family. Her recollection of their tragedies sets off a chain reaction of effects, the first being her taking a decision to confront her husband about injustices towards women in Iraqi society for the first time:

"على الفراش متهيئة، تفتش عن سبب فجاءها الف سبب وسبب. وكأنها تدافع عن المرأة المضطهدة المكممة طيلة حياتها. تتذكر الظلم والقهر والحرمان منتقمة من حال جدتها وجدة جدتها ومعظمهن مُثِّلَ قهرا وكمدا دون ان ينبسن بكلمة اعطتها ما تمكنت عليه في حالة المرأة وهياجها، جوعها واستغلالها وضياع شبابها."

[On the bed, at the ready, she was searching for a reason. And a thousand and one reasons came to her. It was as if she were defending (all) persecuted women, muted throughout their whole lives. Remembering the injustice, tyranny and deprivation and avenging her grandmother and her grandmother's grandmother who had died tyrannised and withered away without uttering a word, she gave to him all that she could about the situation of women, her (sense of) raging anger, her hunger, her exploitation and her youth slipping away].[1]

(1990, 82)

In the Arabic version, we read how her foremothers' histories 'provide' Afaf with "a thousand and one reasons" – an echo of Scheherazade – to link her own personal battle with patriarchy with those stories of women being silenced and forced to suffer injustice in silence for centuries. In English, the political ramifications of Afaf's remembrance goes slightly further: Afaf is presented as a tiger waiting to attack (not defending) and "an advocate": not only to avenge her family foremothers but all (Iraqi) women.

> She sat on the bed like a tiger waiting to pounce, and as she looked for a reason she found a thousand and one reasons coming to her. She was an advocate there to defend women forced to remain mute and under wraps all their lives, there to remember the tyranny and oppression, to avenge her mother, her grandmother, her great-grandmother, all of them withered by the endless injustice without ever uttering a word. And she spoke as an enraged, exploited and angry woman who could see her youth slipping away.

(2005, 84)

In both the Arabic and the English versions, Jalal and Afaf engage in a long argument which culminates in Afaf telling him that all injustices towards women in Iraq are deeply linked to men's capacity for wider political brutality.

> ‏"فإن اجل حدود محسوبة بالكيلومترات او كلمات مثل العرش، الوطن، الدين، القومية يقتتلون فيما بينهم ويلتجئون للسلاح، يخيفون المرأة ، يحرمونها من أولادها ، يزعزعون أمنها واستقرارها . . . ."

> All because of borders measured out in kilometres and words such as 'throne', 'homeland', religion and nationalism, they [men] kill each other, make use of weapons, frighten women by depriving them of their children and destabilise any sense of security and stability.

(1990, 83; 2005, 84)

Her words are more than critical commentary on the concerns of Iraqi intellectuals. Afaf is critiquing the premise of war itself as a destructive politics of ownership and borders enacted by men which frightens women and deprives many of their children, just for stretches of land and ideology. Jalal's reaction is initial incomprehension: ‏"ماذا تقولين؟! الا نحارب من اجل اوطاننا؟!" ["What are you saying?! That we don't fight for our countries?"] (1990, 83; 2005, 84). As her way of thinking beyond borders makes no sense to him, Jalal mocks Afaf, telling her that she should just ask her privileged friend Madeha, the diplomat's wife, whether she would dare to write about her trains of thought that critique both the war in Iraq and the principle of national borders. Instead of being 'deflated' or intimidated by his mocking rebuttal, Afaf draws on the potential power of memory – her differend – once more to give shape to a hope that something *will* be articulated about Iraq in the future – even if this hope is not comprehensible or intelligible at the time that she is expressing it now:

"كفى تهربا, لا تغير الموضوع. سيأتي اليوم الذي يكتب فيه هذا الكلام, باغلب اللغات, اجلا ام عاجلا"

(1990, 83)

> "You're changing the subject. Don't run away. One day someone is going to write about these things in many, many languages. It's bound to happen sooner or later."

(2005, 84)

Afaf's reference to 'many, many languages' uncannily pre-figures the novel's emergence in both languages. Afaf is, in effect, calling on Al-Mana in Arabic – who calls on Charles Lewis in English – to help inscribe and witness the hope that such a record of unknown Iraqi women's stories could come to light on paper even if she does not know when that will be. In wider contexts of Arab women's literature, Hanadi Al-Samman (2010) argues that Arab women's stories can often be read as allegories of resistance and testimonies to the fear of women dying unknown within patriarchal frameworks of gendered oppression. Many stories evoke themes of "enclosure, isolation and non-movement in prisons, rooms, the body itself" (Al-Samman 2010, 85). And while individual Arab women's writings do not represent *all* women's experiences and histories, Al-Samman argues that their existence enacts an important politics of recovery by ensuring that the hope of women's experiences not *going unnoticed* or unread in dominant (patriarchal) epistemes of knowledge remains somehow and somewhere (2010, 85). In Al-Mana's novel, the politics of hope and recovery here creates an uncanny *mise-en-abime* effect (Dällenbach 1989): it is the unknown stories of women calling on 'the house of fiction' (via Afaf) to call on Al-Mana in Arabic and translation to recover their echoes on paper not the writer 'calling on' fictive protagonists.

Particularly radical about the politics of recovery expressed here is that the influence of the Iraqi foremothers does not stop at this point. After remembering her argument with Jalal, Afaf decides to wear a turquoise ring from Iran at a party in London. The ring inspires her and other Iraqis at the party to remember the shared cultural histories between Iraq and Iran, which leads them to be reminded of the thousands of Iraqis of Shi'a descent being deported to Iran during the Iran-Iraq war (1990, 50).The presence of the ring shows that words can be quashed, but other manifestations of (gendered) memory and Iraqi history – such as a woman's ring – cannot. This episode in Arabic was a radical intervention on Al-Mana's part at a time – 1990 – when fear of the Iraqi state surveillance was still high. After this party, we read how Afaf gains the courage to interrogate her uncle – a former Iraqi state employee – about the *other* Iraqis who, at the time, had 'disappeared' in Iraq: political figures killed before and during the Iran-Iraq war. In 1990, this scene is one of the first instances in Iraqi women's literature where individual Iraqi citizens are represented as directly holding (former) agents of the Iraqi Ba'athist state to account, even in the private sphere. It also debunks any myth that all Iraqis were part of Iraqi Ba'athist state ideology during the Iran-Iraq war. Afaf's uncle first reacts to Afaf's questions by avoiding them. Towards the end of the novel, something inspires him to open up to Afaf about his deepest fear: that he will be asked by the state to kill his own son. He weeps as he recalls how Saddam Hussein once 'pardoned' and presented a medal to a father who had shot his own son after he had fled the battlefront – an incident reported to really have occurred in 1982 (Faust 2015, 3). To allay her uncle's tears – and fears –, Afaf responds by reaching out:

"تضع كفيها على جبينه المحزون، تمسده باناملها الرقيقة بنفس اليد التي صفقت ورقصت لاغنية الحرية يوماً متزينة بخاتم فيروزي في زرقته صفاء الكون كله عندما يكون مسالماً وطليقاً مبتهجاً بالحياة. **انها شاهدة وهي معه متضامنة** لم يقتل ولن يقتل ولداً."

[She put her (two) palms on his saddened brow/forehead and massaged it with her delicate fingers, with the same hand which clapped and danced the 'Freedom Song' that day, adorned with a turquoise ring, its blue colour like the whole clarity of the universe when peaceful, free and happy. **She (thus) was a witness and in solidarity with him** that he had not killed and will not kill a son.]

(1990, 133)

She put her hand to his sad brow and massaged it with her delicate fingers. It was the same hand that once clapped and danced the "Freedom song" and had worn a turquoise ring, its blue colour resembling the whole clear universe at peace, free and happy. She was **a witness and she could testify** that he **could** not and **would** not kill any one of his sons.

(2005, 133)

Here we see a ring from Iran playing a role as part of Afaf's expression of care. On Afaf's hand, it is part of what she uses to alleviate the psychic suffering of a long-standing state employee of Iraq – the wordless power of touch. It shows how the veneer of power perpetrated by a system of fear that divided many Iraqis from each other was dissolvable by the wordless power of tears and simple touch, one person at a time. As a text belonging to 'the house of fiction,' the story's 'uncanniness' as a para/text relates to how Afaf's witnessing of two stories is 'witnessed' by 'a (potential) reader' in both languages. The first story concerns Iraqi men of 'fighting' age. Rather than an emblem of Iraqi masculine strength, a soldier is revealed to Afaf as one the most vulnerable groups in Iraqi society during the Iran-Iraq war – liable to be killed by those closest to him upon the command of the Iraqi state apparatus as well as in battle. The second story is how and why Afaf comes to realise that all Iraqis, including a (former) state employee like her uncle, need spaces of safety and comfort in such a climate of insidious fear. In the English version, a third story emerges – but witnessed by the English-language reader only. Afaf presents in English as *consciously aware* of herself as a witness able to publicly testify to the good character of her uncle, an echo towards the retributive frames of justice later brought to bear upon former Iraqi state employees after the overthrow of the Iraqi Ba'athist government in 2003 (Al-Marashi & Keskin 2008). By reading just these few examples of this novel as stories moving across in para/translation, we can read the English 2005 version 're-writes' as working to 're-write' what the 1990 Arabic version could not have written: Iraqis telling stories about Iraqis looking to futures likely to come to be. In 1990, the novel para/translates traces of a future that Al-Mana could only hope for: a hope that her stories – and many other Iraqis' stories – could be recovered from erasure by Iraqi state directives and societal patriarchies. Its 2005 version attests to Al-Mana re/writing the realisation of recovery and hope – via another language – of many diaspora Iraqis' memories. Stories of Iraqis in exile still feeling as connected to Iraq as when they had left.

Daizy Al-Amir: على لائحة الانتظار [*On the Waiting List*] (1988): an uncanny staging

Daizy Al-Amir was born in 1935 and spent part of her early adulthood in Basra, southern Iraq. She passed away in Houston in November 2018 (Abou Rached 2019). She worked as cultural attaché to the Iraqi Embassy in Beirut from 1962 to 1985 and was part of many diverse literary networks. Al-Amir published numerous short stories and seven full short story collections. Representing gendered alienation – or الإغتراب [*al-ightirāb*] – in relation to her location was always a key aspect of her story-telling explained (Al-Amir 1992, 66): memories of Iraq in 1964; Arab society post-1967 war in 1969; increasing scepticism towards pan-Arab ideologies in 1975; confusion, sadness and fear during war in Beirut in 1979 and 1981; ongoing sense of alienation as the wars in Lebanon and Iraq continued in 1988; no sense of belonging anywhere in 1996. In على لائحة الانتظار [*On the Waiting List*] (1988), the theme of transit takes centre stage. Each woman in each story moves between Beirut and Baghdad, occasionally stopping off at unnamed towns, hotels and places in other lands. In some stories, a woman is trying to make sense of her past by relating to objects which signal the presence and importance of unknown others: unsigned letters, photograph albums, uprooted trees. In other stories, a woman seems to be fleeing from someone, somewhere or something only to encounter them in some way again: a friend living through troubled times, an airport, a hotel, an apartment. Habitual itinerance underpins each woman's sense of alienation in each story, meaning that perpetual uneasiness functions as a default 'normal.'

This story collection in English raises a number of questions relating to its own construction as a para/translation. From a perspective of 'feminist paratranslation,' the outer cover jacket appears as if it is working to help the reader understand the numerous frames of reference framing its production. The large iconic image of The Corniche, Beirut, depicts the city where Al-Amir spent two decades of her life (Al-Amir 1989/1994, x): Beirut. A small paragraph on the back outer jacket cover explains that the Lebanese Civil War (not war in Iraq) is the stories' political backdrop. Its title – *The Waiting List: An Iraqi Woman's Tales of Alienation* – suggests that the collection is a 'fractured' novella – a short novel made up of disconnected chapters all told by one Iraqi woman. As an academic publication from the Centre for Middle Eastern Studies, University of Texas, this work also presents very differently to Al-Mana's self-published and self-translated novel. All of its copy editing is complete. The translation by Barbara Parmenter is flawless. The etching-like illustrations for each chapter communicate what each story is about. In a concise and detailed introduction, US-based Arab literature expert Mona Mikhail (1994) critically situates the importance of Al-Amir's writing within contemporary Iraqi literature and women's literary activism in the Arab world in general. The short editorial introduction by Annes McCann-Baker (1994, vii) explains who was involved in this "transnational" publication as an "international collaboration of effort," starting from the outer cover photographer, the chapter illustrator and the US-based women academics who brought Al-Amir to meet McCann-Baker in the United States. Barbara Parmenter is introduced as the "expert Arabic translator" and Mona Mikhail as the academic expert, and Al-Amir is credited with helping with the translation and image selections.

There are nonetheless interesting unexplained silences in this para/translation. McCann-Baker recalls that Al-Amir's 1988 collection was meant be translated "as quickly as possible" after meeting her in 1989 (1994, vii). The book, however, did not get published until 1994. Al-Amir's "Author's preface' is divided in two parts, the first (1994, ix–xiii) ending with "Baghdad, Autumn 1989" (xiii), the second with "Beirut, March 1994." Neither Al-Amir nor McCann-Baker explain the delay. Another silence relates to authorship. McCann-Baker clearly situates Al-Amir as an Iraqi woman writer presenting 'her' story to new audiences, including helping with the English translation. A brief glance at the ISBN page reveals that the book is cited as "by the Center for Middle Eastern Studies at the University of Texas in Austin" (1994, iv), not Al-Amir (or Parmenter). Another deafening silence of that of Al-Amir's political positionality towards the wars in Iraq. In her 1989 preface, Al-Amir recalls that she left Beirut in 1985 to return to Iraq, when the Iran-Iraq war was as its height. Despite being an acclaimed writer inside and outside of Iraq, Al-Amir explains in her preface that she found herself "unable to involve [her]self . . . to write fiction about it" (1989/1994, xii) – no easy feat in Iraq when silence was not an option for many Iraqi writers living in Iraq (Wali 2007, 52). Her 1988 collection على لائحة الانتظار [*On the Waiting List*], as a collection of stories dedicated to specifically individual and private experiences of women, was clearly a notable exception to other stories circulating at a time when Iraqi literary and aesthetic production was dominated by depicting the glories of the 1980–1988 war (Davies 2005; Rohde 2010; Khoury 2013). Al-Amir seems to have deftly situated her silence on the war on Iraq as a result of being a 'new' Iraqi returnee from Lebanon impacted more by the war and her sense of الإغتراب [*al-ightirāb*] there. In the 1994 preface written in Beirut, Al-Amir refers to how the 1990–1991 war in Iraq obliged her to stay in the United States and travel to Lebanon but not Iraq in 1994 – the land of her mother, rather than her father. Her return to Lebanon however did not seem to give her a sense of peace as the preface ends with Al-Amir (1994, xiv) expressing how uncertainty about her future is "terrifying my soul," without her giving any further detail on what exactly this terror is.

While the reasons for all of these 'gaps' in the English para/translation cannot be known at this point, one way of exploring them further is by looking at how Al-Amir is para/translated in other locations of the book. In her introduction, US-based Iraqi academic Mona Mikhail (1994) presents Al-Amir to English language readerships first with a detailed overview of Iraqi literature up until 1967. Although Al-Amir, along with many Iraqi writers, was clearly publishing stories after 1967, Mikhail makes no detailed reference to any other Iraqi literature published post-1967. She does, however, situate Al-Amir's work as part of a tradition of Arab writing which reflected "the aftermath of the 1967 setback" (1994, 3), that is, the 1967 war between Israel and neighbouring Arab states, by "forcefully expressing disenchantment with the ruling regimes and the endemic lack of freedom in societies" (ibid). Mikhail does not refer to the Iraqi Ba'athist government categorically assuming power in Iraq in 1968 as part of this disenchantment, nor its potential impact on the writing of Al-Amir and other writers in Iraq. Instead, she draws on the words of Syrian woman novelist Ghada Al-Samman to illustrate why Arab women's writing, including Al-Amir's, expresses political disenchantment through the lens of gender:

> As an Arab citizen, a woman suffers from all the constraints imposed on any of her compatriots . . . in addition the attempt by women to restore their rights is part of the attempt by Arab individuals to restore their very humanity (*al-Qabila tastajwib al-Qatila*, Ghada Samaan, 1981).
>
> (cf. Mikhail 1994, 4)

Here we see a chain of presentation: Mikhail is presenting how Al-Samman presents the power dynamics of repression towards women in the Middle East. Al-Amir alludes to the politics of her writing being gendered in similar ways in her 1989 preface: "The main character was always a woman because I understand women more than men, although they certainly play their roles to perfection on life's stage – men with false bravado and women with chronic fear" (Al-Amir 1989/1994, xii). Al-Amir does not clarify what this 'stage' of life is, and Mikhail makes no reference to Al-Amir's comments. Al-Samman's commentary on Arab women's writing seems 'staged' by Mikhail to reflect how Al-Amir may have wanted her stories in English to be understood by new readerships: as stories about one women's sense of gendered alienation readable in multivalent ways, with no direct comment on her politics.

This is not to say that Al-Amir is *not* mediated as 'communicating' with her new readerships herself. Quite the reverse: Mikhail presents Al-Amir's stylistics of writing as directly interactive with all of her readerships: "Whether she uses the narrative 'I' or the conventional third person, Al-Amir is transparently narrating her present itinerant reality" (Mikhail 1994, 5). Mikhail juxtaposes stories in the collection with associations of Arab world folklore alongside more contemporary points of reference: "A certain minimalism and terseness in her style speaks directly and convincingly to a hurried and on-the-go urbanised reader who is probably also on a 'waiting list' at a Wag Wag airport" (1994, 5). "Wag Wag Airport" is the title of one of the chapters which encapsulates the sense of alienation that Al-Amir evokes in this collection, but Mikhail does not explain it further. In brief, 'Waq Waq Island' refers to a mythical island with trees bearing fruit that look like naked women making the sound 'waq waq' (Malti-Douglas 1991, 7). During the Islamic Empire era, 'Wag Wag Island' represented an "inverted world" (Kruk 1993, 226) where warrior women ruled and enforced law and men carried out "handiwork" (ibid,

216). Believed to be in the Far East in reality, this island became synony-mous in former Islamic Empire folklore with the "conceptual limit of the known world" (Toorawa 2007, 57). The associations of 'Wag Wag Island' in Arabic folklore are an important paratext of this work, as the story "Wag Wag Airport" in Al-Amir's collection is indeed inverted, but in somewhat 'off-tilt' ways – it is about two women trying to find their way out of an airport where passengers are sent from gate to gate by airport staff who stand by, misdirecting and mocking them rather than helping them. For an Arabic-language reader, this story potentially evokes many questions: what inversion is this story referring to? Who is making the 'waq waq' sound? In which 'unknown' world is this airport? While the trace of this story's importance is alluded to by Mikhail in her introduction, an English-language reader is left to consider what the para/text of 'wag wag' could possibly mean. In this way, Mikhail presents Al-Amir as doing what she does directly in Arabic: leaving uncanny poetic traces of confusion.

While Al-Amir is para/presented by Mikhail and McCann-Baker as speaking to readerships directly, the voice of the translator, Barbara Par-menter, is notably absent. Concerning the ISBN page, the work is moreover also presented as authored by the publisher, not by Al-Amir or the translator. Gayatri Spivak (1993, 183) warns of how neo/colonial global markets mis/shape and overwrite writers – and their rights to authorship – from certain parts of the world. From a perspective of feminist paratranslation, I ask whether this is the case with this collection. At the time of its publication, Nancy Gallagher (1995) read the English version as a deeply personal act of one woman's "cathartic" writing "with politics a distant backdrop" (1995, 63). In later years, Shamanez Bano (2015) praises Al-Amir for "sensitively" describing "the impact of the rise of Saddam Hussein as an Iraqi leader" (2015, 4), although not one Iraqi political figure is mentioned in this col-lection at all. These brief reviews illustrate how each critic interprets the silences of this work by drawing on their own epitexts of understanding on the writer and Iraq. They also show that Daizy Al-Amir is read by both reviewers as the aut/her of this story. One way of considering this gap, or 'differend' is that of its times of publication. The English version is some-thing that its Arabic version would never have been in 1988 when published: a story of two uncannily failed returns. In 1988, the Arabic version was a collection of stories by an established Iraqi woman writer experiencing the pain of a one-way return to Iraq after decades of life in Lebanon. The 1994 is a book of *two* journeys away from Iraq – to the United States in 1989 and then again to Lebanon in 1994, as signalled in the two prefaces by Al-Amir (Al-Amir 1989/1994). It is the prospect of the second or failed return to Iraq which inspires within Al-Amir her fear, the reasons for which are never explained. Was the 'publication' or 'rights' of 'the book' then assigned to a *de*personalised agency – such as a university press – to protect the book – and Daizy Al-Amir – from any repercussions should she have to return to Iraq? Or is the ISBN a recognition of the co-collaborations bringing this work together? Or something else? Such questions bring into stark relief the fact that many attempts to identify 'differends' in a para/translated work rarely 'result' in definitive 'answers' concerning a work's meaning-making. Asking such questions is nonetheless important as a way of considering 'opacity' and absences in para/translation as an integral part of a translated work's critical scope.

Conclusion: new paradigms of c/overt translator agency

I have focused on how 'silent' translator presence emerges as an integral aspect of حبل السرة [*The Umbilical Cord*] (1990) by Samira Al-Mana and Daizy Al-Amir's على لائحة الانتظار [*On the Waiting List*] (1988). Al-Mana's stories enact a politics of gendered recovery from Iraqi state censorship twice, Al-Amir's (potential) double attempt at flight away from it. If we consider borders between text and paratext, translation and paratranslation, translator and paratranslation as non-definitive we can fruitfully explore c/overt or liminal manifestations of para/translator agency as important sites of activity. Appreciating that much of both writers' meaning-making in these two stories was written to be concealed – yet alluded to – in both languages provides valuable insights into what we think we are reading when engaging with a translated work. This point enhances appreciation of the many gaps we may 'find' (or not find) in earlier examples of Iraqi women's literature in translation as part of their meaning-making moving across languages. This particular point helps us recognise the complexities that some writers – and (para)translators – may have faced when mediating opaque or uncanny expressions of 'ambivalence' in 'the house of fiction' (Bhabha 1994). It also helps us read silences differently. Reading both works as 'uncanny' para/translations sharpens the focus on where different representations of ambivalence can be read in a para/translated work. It also diversifies the premise of 'one' translator as the principle site – and agent – of activism.

Note

1 All back translations cited between square brackets are my interpretations of the Arabic texts.

3 Translating gendered dis/location in post-2003 Iraq

Inaam Kachachi

The theme of location and Iraq – geographical, social, political and corporeal –
has inspired multiple forms of aesthetic activism by Iraqi artists and writers who work to challenge hegemonic power relations and political oppression (Al-Ali & Al-Najjar 2013). Much activist work by Iraqis concerns the
impact of post-2003 US occupationv of Iraq on aesthetic expressions of
Iraqi identity. Iraqi writer Ali Badr (2013) observes how Iraqi artist/writer
identity after 2003 became binarily interlinked to "location" (2013, 115),
with some Iraqi writers still living 'inside' Iraq as taking the following position: "the internal literature is the only true one; as those living overseas,
these are the cowards, we have lived through hell for many years" (2013, 117)
and those 'outside' Iraq responding: "you lived under a dictatorship, the
authority destroyed your perspective, making you incapable of producing
a true and humanistic literature, because there was no escape from censorship" (ibid). Intersecting with these debates are concerns about the effect of
US interest on Iraqi aesthetic production in the 2003 war in Iraq. Iraqi art
critic Nada Shabout warned that:

> within today's interest in all things Iraqi, Western media has taken the
> liberty to define Iraqi art and publicise the image it found fit for the
> world's perception of what this art should look like. In other words,
> Western media is "inventing" a new historical narrative for modern
> Iraqi art.
>
> (Shabout 2013, 7)

As the *perceived* location of post-2003 Iraqi cultural production as well as
the *Iraqi* artist/writer her/himself emerged as an integral part of a work's
political agency and meaning in different ways soon after the 2003 war, I
explore the 'politics of re-location' intertwining الحفيدة الأميركية [*The American Granddaughter*] (2009a; 2011) by Inaam Kachachi written in Arabic in
2009. It is the first novel by an Iraqi woman writer translated into English to

focus on the politics of Iraqi women translating during the 2003 war in Iraq. In this chapter, I focus on the politics of different translator 'voices' moving across languages using 'audible' paradigms of feminist translation analysis.

The Arabic version of the novel has been translated into English by two different translators: one 'official' translation by Nariman Youssef published by Bloomsbury Qatar Foundation Publishing (2011) and an online version by William Hutchins (Kachachi 2009b). It is a novel about US-Iraqi inter-preter Zeina, who works with the US army in Iraq to help liberate her people from Saddam Hussein (2011, 10) and her sense of conflict towards her North American, Iraqi and exile identity. She speaks, writes, listens and translates formal written Arabic, Mosuli Iraqi spoken Arabic and US English within the hegemonic masculinised frames of war and oppression in Iraq and seems trusted by no one as a result. To explore how the gendered politics of all three languages in the Arabic version can be read or 'heard' when mov-ing across into English, I draw on two analytical approaches of gender-conscious translation which have not so far been brought alongside Iraqi and Arab women's literature in English translation: Helen Kolias' approach of "listening translation" (1990) and praxes of "mimetic translation" used by feminist translators in Quebec (Flotow 2004, 93). Going beyond the language in question, these analytical frameworks of feminist translation analysis focus on the audibility of translation and how the politics of orality and interventions on the part of the translator can be 'read' in translation. Such perspectives, in my view, are helpful starting points when reading how two different English translators appear to 'hear' Inaam Kachachi's Arabic gendered critique in different ways and so respond to them differently.

الحفيدة الأميركية [*The American Granddaughter*] (2009): critical contexts

Born in the early 1950s, Kachachi grew up in Iraq when tensions between the Iraqi Communist Party (ICP) and the Iraqi Ba'athist Party were extremely high, with essentialised notions of 'Iraqiness' becoming ever more prevalent within political echelons of Iraq from the 1960s and 1970s (Ismael 2007). In 1979, Kachachi left Iraq to study in France due to the stifling nature of the Ba'athist rule in Iraq (AbdelRahman 2012, 1) and later worked there as a journalist. Her first book, *Lorna, Her Years with Jawad Salim* (1998), was a biography of the wife of Jawad Salim, an Iraqi sculptor famous for the sculpture titled نصب الحرية [*The Monument of Freedom*], a huge monu-ment in Baghdad celebrating the Iraqi people and 1958 Revolution, an event which occurred with strong support from the Iraqi leftist movement (Ismael 2007). Kachachi also made a documentary about Naziha Al-Dulaimi, the first woman minister in an Arab country with that same 1958 Revolution government. Within the contexts of pre-2003 Iraq the subject matter of Kachachi's earlier works represented a clear resistance to Iraqi state directives which sought to erase traces of the political left as part of Iraq's 'official' history. As she never returned to live in Iraq, Kachachi's sense of Iraqi identity is something that she configures as inside herself and the Iraqi peoples themselves rather than within political distinctions imposed by hegemonic practices of power (Lynx-Qualey 2014). Like many of her works, Kachachi's الحفيدة الاميركية [*The American Granddaughter*] (2009a) challenges essentialist discourses of nationhood and Iraqiness via the prism of gendered language and the politics of Arabic–English translation during the 2003 war in Iraq.

الحفيدة الأميركية [*The American Granddaughter*] centres on Zeina, an Iraqi-American woman translator working in the US army, her Mosuli Iraqi grandmother and the different politics of their shared connection to a troubled post-2003 Iraq. Zeina writes a diary which documents her journey to the United States to Iraq and back again, noting her changing relationships to her Iraqi grandmother, fellow Iraqis and the United States in the process. While writing her memoirs, the fear of an editing, censoring Iraqi woman author haunts Zeina (2011, 26), which leads her to consider her troubled relationship with her memoir-writing as a contest, a struggle for a platform to speak and a battle for identity and political agency. At the end of the story, Zeina writes that she has killed off the Iraqi woman writer "before she could kill me" (2011, 178). Kachachi's story, however, suggests that people's identities are more fluid, self-reflexive and transient than Zeina believes and are mostly influenced by lived experience and instances of reflection. Zeina's experience, for example, is shown as marginalisation in the United States, violence in Iraq and disillusionment with both. Kachachi also shows how some notions of identity can be negotiated despite all the prevailing frameworks of power implying otherwise, as Zeina believes. Zeina's Iraqi grandmother speaks the Mosuli Iraqi dialect, which is then transcribed by Zeina in written formal Arabic – and later English translation – into her diary. Formal written Arabic is certainly a language of power which frames all 'spoken' or 'vernacular' Arabic as out of place in the Arab public sphere (Safouan 2007). Yet by Zeina writing how her grandmother and her family actually speak into the 'written' fabric of her diary, women and men who do not read and write are conversely given a platform to 'speak' to readers via this language of authority which would otherwise overwrite them or configure them as absent – para/translated by Zeina, Kachachi and the very act of writing itself. The leitmotif of written Arabic as a textual cipher which 'translates' how Iraqis 'speak' and experience their political realities is a fundamental aspect of this novel and the questions of identity it poses.

Alongside its interrogations of aut/her/ships via different languages, this novel makes many pointed critiques on essentialised configurations of political affiliation. When Zeina arrives in Iraq in 2003 as an interpreter, she recalls Iraqis looking at her as "simultaneously their daughter and their enemy while they could be my kin as well as my enemy" (2011, 7) as she sits beside US soldiers in military vehicles. While Zeina believes that her identity is by no means fixed, she accepts that fixed ideas of identity are realities that she has to deal with, often through التقمّص [*al-taqammuṣ*], meaning literally 'changing shirts.' To visit her Iraqi relatives, she wears a black *abaya* (2011, 72) to mark herself as Iraqi and as a woman. As an American citizen in the US army, she has to don "something like masculinity" (2011, 143) – the US military helmet and sunglasses. The theme of التقمّص [*al-taqammuṣ*] haunting Zeina are shown as relating to the post-9/11 attacks in New York: by enlisting in the US army she thinks she can repay a debt of gratitude to the United States (for being hosted in exile) *and* prove that she and her family are not one of 'them' – 9/11 terrorists. Fadwa AbdelRahman (2012) describes Zeina's 'becoming' a US army interpreter as a permutation of "the simultaneity of conspicuous togetherness and conspicuous otherness in a predominantly globalized world" (2012, 2) in post-9/11 America, or what Gloria Anzaldúa (1987, 3) terms "una herida abierta (an open wound) where the third world grates against the first and bleeds." In many ways, this novel represents more than the lived realities of 'donning' identities in charged situations. It represents the hope of something of Iraqi identity being salvaged in the face of destruction. When her grandmother dies – after seeing her granddaughter wearing a US army uniform – Zeina decides to leave Iraq. Her comfort is, however, an oath never to forget Baghdad (2011, 180). She returns to the United States no longer a granddaughter nor feeling Iraqi or American, her oath leaving a trace that she could remain all three.

The act and status of translation are extremely charged 'sites' of identity in this novel, and the consequences of intermixing Arabic and English in post-2003 Iraq are often deadly. And Zeina hides her identity for her and her US military unit's safety (2011, 6) for good reason – a local Iraqi woman working with the Americans is later found with her throat slit and her eyes gouged out (2011, 86). Zeina later discovers that differing understandings of military service across borders make her situation as an interpreter vulnerable. As military service in Iraq was obligatory, the actions of US male soldiers in the eyes of some Iraqis were assumed as linked to a chain of command they had little say over. The agency of Iraqi interpreters, however, was seen as more treacherous, as this role is seen as voluntary (2011, 144). As an interpreter, Zeina finds that her position for US army colleagues is not much easier. She is often asked "whose side are you on?" (2011, 106).

Inaccurate cross-border discourses moving in the international political sphere are shown as disastrous and going beyond the language in question:

> We [the US] sold you a dream that was too good to be true. But we weren't alone. You had your spin doctors and nuclear scientists and generals. They told us about weapons of mass destruction. . . . September 11th was waiting for a scapegoat, so we bought it all. You believed us, and we believed them.

(2011, 164)

Despite the novel's charged subject matter, however, little information on the politics of either of the two English translations is provided. In the 2011 book version translated by Nariman Youssef, the novel's IPAF 2009 Shortlist status is emblazoned on the front cover. A glossary of Arabic and Iraqi cultural words alongside the brief bio on the translator Nariman Youssef show the book's status clearly as a translation but explain nothing of its translation processes. The 2009 online version – three chapters of the book – by William Hutchins does not explain anything about its contexts of translation either, other than that he has copyright to his English translation via the French version of the novel published by Éditions Liana Levi in 2010. While such gaps raise questions on how to read the para/translatory politics of Kachachi's book, other fundamental questions about الحفيدة الأميركية [*The American Granddaughter*] concern this novel as a story about the politics of translation across different registers and languages. How did each translator read such politics of translation as represented by Kachachi in Arabic?

As the novel's different registers of Arabic *are* indeed part of the novel's diffuse political critiques of essentialised notions of identity, the question is then how to read two potentially different translator engagements with language register if no information has been given on their processes of translation. Helen Kolias' paradigm of 'listening' translation raises useful thoughts about how we could work to read how the story's registers move across languages with and without para/textual explanation. According to Kolias, 'listening translation' involves a translator *listening* to the cadences s/he reads in the 'source' text and *attesting* to her/his own experience of listening to and reading the 'target' text. This attestation can be shown in different ways. In her own contexts, Kolias (1990, 214) was translating what she termed as a 'minor' text (nineteenth-century Greek woman writing in a Greek island dialect) into a 'major' language (English). To explain what a 'listening translation' praxis was for her, Kolias (1990) gives practical examples of how translators can "attest to what they hear, and thus leave their mark as co-creators of a space for a 'foreign' text" (1990, 219). One way is by a translator making use of prefaces to show how her/his own listening impacts how she has translated (1990, 217). Another way is by a translator showing the *process* of listening throughout the text by using foot-notes (ibid). In contexts of feminist translation studies, listening approaches to translation focus on practices of mediating texts to be experienced across different mediums, such as spoken poetry or theatre for more political purposes. In 1990s Quebec cultural scenes, a 'mimetic' or 'ventriloquist' approach resulted in some translators working to render 'the sounds' of a source text rather than its semantic meaning to communicate an *experience* of language rather than a simple surface rendering of words (Flotow 2004, 93). Politically, such mimetic approaches also worked to bypass patriarchal lexical language structures that flatten or suppress voices of resistance by 'rewriting the echoes' of what cannot be represented in (masculinist-based) language to be perhaps heard or audible later. Barbara Godard conceptu-alised such traces as "an echo of the self and the other, a movement into alterity" (Godard 1989, 44) which may not be audible in all contexts but can nonetheless be configured as an act of "transformance" (Godard 1989, 46) – a performance of transformation of an inaudible echo. As neither 'listening' or approaches have been explored in contexts of Iraqi or Arab women's literature, I explore how the 'audible' politics of this novel moves across Arabic and English translation from both perspectives. As Kacha-chi's novel has been translated into English (three chapters) twice, I first read the chapters translated by both Nariman Youssef (Kachachi 2011) and William Hutchins (Kachachi 2009b) to 'listen to' how Inaam Kachachi's gendered critique of war in Iraq is 'heard' differently by the two translators. I then explore how Youssef's version re/locates the 'echoes' of conflicting gendered authorships at the crux of Kachachi's novel as an Arabic-language work – which is to reconfigure Iraq and Iraqi identity as prevailing in its peoples despite the wars and conflict scattering them inside and outside of Iraq in different locations.

الحفيدة الأميركية [*The American Granddaughter*]: 'doubled' in translation

As explained previously, *The American Granddaughter* (2011) was pub-lished by Bloomsbury-Qatar Publishing Foundation after the Arabic ver-sion of the novel had been shortlisted for the 2009 (IPAF). Prior to the 2011 version, the French version, titled *Si je t'oublie, Bagdad* [*If I Forget You, Baghdad*] (Kachachi 2010), was published by Éditions Liana Levi. *France Culture* praised the French version as "le double portrait d'une femme et d'un pays déchirés" [the twin portrait of a woman and a country ripped apart].[1] A 2013 blog describes how "la narratrice subit un dédoublement de person-nalité: d'un côté la soldate qui défend son pays; de l'autre la romancière qui capte chaque image, chaque souvenir de sa grand-mère pour un potentiel roman quand elle rentrera aux États-Unis" [The narrator undergoes a split-ting doubling of personality: on one side, she is the soldier defending her country; on the other, the novelist capturing each image, each memory of her grandmother for a potential novel to be written when she returns to the United States].[2] The motif of "un dédoublement" or 'doubling' is an insight-ful metaphor as it reflects something of the ambiguous sense of boundaries in a globalised war in multiple locations inside and outside Iraq and the United States. In the Arabic version, the title الحفيدة الأميركية [*The American Granddaughter*] echoes Zeina's three identities assigned to her before trav-elling to 'perform' the role of interpreter for the US army in Iraq: Arab world gendered generational kinship ties, Mosuli Iraqiness and American-ness. Directly following the title before its first page is a Qur'anic saying:

[Beware of the tender green (plants) (growing) from animal excrement]

إياكم وخضراء الدمن

[Prophet *hadith* (whose authenticity) is not agreed upon]

حديث نبويّ غير متفق عليه

What we read here is a *hadith*, a saying attributed to the Prophet Muham-mad, which exhorts men to avoid associating with a woman who appears beautiful but whose background is of ill-repute (Al Halaby 1999, 13). It implies that a person's origin – and whatever 'nourishes' or creates a person – is an indication of their worth and/or moral behaviour. Read after the book's title الحفيدة الاميركية [*The American Granddaughter*], this epigraph intimates a warning about this 'American granddaughter': coming to Iraq under the auspices of wanting to help her first homeland, nourished and created by the US military. As the novel goes on, it becomes clearer that this saying could refer to many other political motifs which go far beyond gender itself: new political ideologies for Iraq presenting as 'new' post-2003 being dubious in origin; the US army presenting itself as the harbinger of a 'new' democracy in Iraq, its intervention of liberation mired in filth – political self-interest. It also serves as a neat metaphor for critiquing 'new' forms of violence emerg-ing in a post-2003 Iraq. At the end of the novel, Zeina throws her army kit out with the garbage, writing "I wouldn't be planting basil in my helmet. Sweet perfume doesn't grow in metal. That's what I write to Muhayman [a relative fighting against the United States in Iraq] but he doesn't reply" (2011, 179). By this she means what grows from violence – however ini-tially appealing in appearance – will never provide healthy nourishment. Another layer to this epigraph is its very status as a weak, disputed Prophetic saying (Al-Halaby 1999, 13). In the Arabic version, the words غير متفق عليه, literally meaning "not agreed upon," show that many Arab world religious traditions can be subject to critique and disagreement as any other saying or form of knowledge. With the politics of sectarianism showing itself to be lethal for many Iraqis (Zangana 2014), Kachachi's use of a weak Islamic *hadith* in a novel about conflicted and shifting loyalties in post-2003 Iraq emerges as politically ambiguous, if not blasphemous (at least to some Muslim readers). Or it could be read as a commentary on the politics of writing and the authority to speak. Such doubts echo, double back and reverberate throughout the novel in different ways ز According to critic Faruq Youssef (2010), the ambiguity of aut/hership thread through this novel is why both Kachachi and Zeina gain some respite from their shame in the part they have to play in reiterating the 2003 politics of war in Iraq as Iraqi writers and 'translators' co-opted into the very hegemonic structures they both critique.

How can we read how such multilayered discourses echoing in the English translations? The 2011 version translates the Arabic title as *The American Granddaughter*. The outer cover shows the upper face and eyes of a young dark-haired woman merged with a sepia landscape of palm trees in the desert. The line "Shortlisted for the 2009 International Prize for Arabic Fiction" followed by "Inaam Kachachi" appears with no mention of the translator, Nariman Youssef. The *hadith* epigraph features as follows:

> *Beware the beautiful woman of dubious descent* – an unauthenticated Hadith

The translation is an explanatory translation of the *hadith*. There is no reference to green plants growing from animal excrement. The subtle political reverberations of critique towards the war in the title and epigraph are muffled, positing the gendered epigraph and feminine appearance of Zeina as the *hadith*'s point of departure as it simply refers to women. It 'cuts out,' for example, the echo resonating across Zeina's later refusal to grow basil in her helmet towards the end of the book (2011, 179). In his 2009 online version of three chapters of the book, Hutchins does not translate the epigraph at all. He does make an interesting intervention with its title that reverberates the multilingual dimension of this novel as a translated work moving out of and across different languages in international publication. Hutchins translates the title as *If I Forget You Baghdad*, rather than as *The American Granddaughter*, an intervention which connects his version to the French version's title, *Si je t'oublie Bagdad* (2010). This French version's title does not come about by chance: it is from the novel's final line: "شُلَّت يميني اذا نسيتك يا بغداد" [May my right (hand) be paralysed, should I forget you, Baghdad] (2009, 196; 2011, 180) or "I'd give my right hand if I should ever forget you, Baghdad" (2011, 180). This turn of phrase is an adaptation of

Psalm 137.5[3] from the Old Testament and shows Zeina invoking her Iraqi Christian identity at the end of novel alongside the echo of Jewish diaspora exile from Jerusalem by Nebuchadnezzar, the king of the Chaldeans in BC 597/586 (VanderKam 1985, 126; Gillingham 2013, 65). It is at this point that Zeina remembers her father, a Chaldean Christian Iraqi, invoking this Psalm, changing 'Jerusalem' to 'Baghdad.' This transformation of a Psalm about Jerusalem offers a trace to how alterity can be 'reclothed' in translation in different ways and across different times: the one-time Chaldean historical oppressors echoing in the first version of the Psalm now identifying as the oppressed. Hutchins' echo of the French title of the novel foregrounds two aspects of the novel less apparent in the title *The American Granddaughter*: Zeina's identity as a Christian Iraqi and her decision to safeguard her Iraqi identity through the act of memory.

A comparative reading: the politics of gender (dis/re)location in re/translation

An initial reading of the chapters translated by both William Hutchins (Kachachi 2009b) and Nariman Youssef (Kachachi 2011) show that both translators have an excellent grasp of Arabic and English. Yet both translators seem to 'read' or 'hear' Zeina's voice slightly differently. A comparison of examples by the two translators is a useful way of highlighting the influence of translator 'listenings' on what we read in translation.

"إحتشدت الكلمات في رأسي وتسارعت وتدافعت وتداخلت مثل غيوم بيض تهرب على عجل, ثم توقفت مرّة واحدة وزخّت **مطرها الحاذق** على اصابعي . . ."

[The words were crowded in my head and sped and dashed off to move into one another like white clouds fleeing in haste to then stop all at once and pour down their bitter rain on my fingers.] emphasis added.

(Kachachi 2009a, 34)

Words clump together in my head and then quickly speed off only to dash against each other before blending into one another, like white clouds that immediately flee. Then these words pause once more and pour through my fingers like **a deluge**. [emphasis added].

(Hutchins in Kachachi 2009b)

The words filling my head are white clouds taking flight. They move and merge and change shapes, then all at once they stop and pour forth **their acid rain**. [emphasis added].

(Youssef in Kachachi 2011, 26)

Hutchins (Kachachi 2009b) has tried to capture Kachachi's use of reflexive verbs by his inclusion of expressions such as "each other" and "one another." Youssef's sentences are half the length of Hutchins' and appear to follow the more succinct style of Kachachi's Arabic. Youssef's choice of "acid rain" (Kachachi 2011, 26) for Kachachi's metaphor of 'sour' resonates with Zeina's metaphors of writing as an act with malignant effects. Hutchins chooses "deluge" to transmit Zeina's writing while in Iraq as proliferation rather than as an act with bitter or acidic quality. The purpose of reading showing these two examples here is not to judge which translation is 'better' than the other. What I do want to show is how two different 'hearings' of the Arabic versions can have very significant impact on the readings of this novel in English. Such hearings are important and necessarily charged due to the political nature of the novel itself. One particular emblematic example is in the chapter when Zeina is describing her transformation from living as a civilian to becoming one of many US military personnel during her journey to Baghdad. This is a chapter where Kachachi works, in Arabic, to highlight and parody Orientalist discourses of women and war during the post-2003 Iraqi/US military intervention. During her journey to Baghdad from the US, Zeina recalls an Egyptian interpreter, Nadia, whom she liked very much and who seemed very streetwise. Zeina describes Nadia as "محتالة بالفطرة" [*muḥtāla b'il-fiṭra*] (2009, 31) as she artfully commanded the attention of the US soldiers in their unit by her lively way of speaking. In Arabic, محتالة [*muḥtāla*] means "artful, cunning" (Wehr 1994, 255) and بالفطرة [*b'il-fuṭra*] means "by nature, in disposition" (Wehr 1994, 842). "محتالة بالفطرة" [*muḥtāla b'il-fiṭra*] has moral, playful but not necessarily sexual connotations and alludes to dynamics of sexuality consciously exploited by Nadia in hyper-masculine US army settings. Youssef uses the North American phrase "hustler by nature" (2011, 23). Hutchins, however, translates محتالة بالفطرة [*muḥtāla b'il-fiṭra*] as "a born prostitute" (Hutchins 2009), a choice which raises many questions, the first being: what can the term 'a born prostitute' mean? While the sex industry, a world-wide phenomenon, often intermeshes with endemic poverty, trafficking and other human rights violations (Mattar 2011), no one is ever 'born' a prostitute. By situating this term within somewhat essentialising frames of sex and economy, Hutchins *dis*locates the Arabic register and adds a new trace of gendered derogation to this English version of the story.

Later in this scene, Zeina does indeed situate the dynamics of Middle East politics as linked to wider post-2003 Orientalist sexual imaginaries, but not in the way that Hutchins translates محتالة بالفطرة [*muḥtāla b'il-fiṭra*]. Rather, Kachachi seems to be making a point about the politics of neo/coloniality:

"كل ما حولنا خشن وذكوريّ، ونحن غير مدرّبات بعد, على الاسترجال. لا ينفع هنا، الجهاد للحفاظ على الأنوثة. انت إمّا جنديّ أو جارية ."

[Everything around us was rough and (hyper) masculine and we were not trained up (pl. fem) yet to act like men. It was of no use here to do jihad for femininity. You are either soldier or slave-girl.]

(Kachachi 2009a, 31)

Tropes of gender consciously abound throughout this excerpt in multiple ways, parodying the binaries of neo/Orientalism relating to Iraq, the United States and 'the rest.' The US army is shown as hyper-masculine and the interpreters as "not-trained-up" (ibid) to deal with its military roughness. Conversely, "femininity" is juxtaposed with "الجهاد" [al-jihād], a word associated with (masculinised) Islamic holy war. Here Kachachi uses this word to mean a potential fight against the onslaught of US hyper-masculinity. Any attempt to "do jihad" for their version of femininity would, however, according to Zeina, 'locate' the women interpreters in the US army within Orientalist tropes of Arab women – as "slave-girls," spoils of the war and having no personal or symbolic agency. This passage shows Arab women interpreters unable at this point to transcend a pre-Orientalised neocolonial location. Their options seem starkly binary – you either become 'one of us' or 'one of them,' an echo of post-2001 US-state political discourses by US President George W. Bush, who described the war on terror as 'a crusade' and a test of political loyalty to the United States and its values as a global military power (Bush 2001). Zeina shows her growing awareness of women interpreters going through a process of *becoming* native informant-turned-military-man – or Orientalised spoil of war but remaining a participant and bystander, making no effort to turn back. Her phrase in some ways echoes what Trinh Minh-Ha termed as the *triple*-bind facing many women (writers): viewed through being women, colonised/occupied and racialised in neocolonial contexts but writing anyway: "She knows she is being different while at the same time being Him. Not quite the Same, not quite the Other, she stands in an undetermined threshold place where she constantly drifts in and out" (1995, 218). If we keep Minh-Ha's framework of a triple bind in mind, Zeina seems to be writing her apparent assimilation into the US army as a choice to be parodied rather than justified: awareness of herself as not the Same as the US army but not Other to it, as she has made a conscious choice to situate herself within its machinations. In English, Zeina's (self-)parody and ambivalent assimilation is heard by Hutchins as:

Everything around us was rough and masculine but we hadn't been trained yet to act like men. It was pointless here to attempt to hang onto your femininity. You were either a soldier or **a call girl**.

(Kachachi 2009b, emphasis added)

Youssef attests to what she hears by using more overtly gendered and Orientalist terms of engagement:

> Everything around us was rough and masculine and we weren't yet trained to be **macho**. But it wouldn't do to cling on to femininity. Here you were either a soldier or **a concubine**.
>
> (Kachachi 2011, 25, emphasis added)

While both versions are very similar, the subtle differences between them are important. Hutchins (2009) translates "الجهاد" [*al-jihād*] as "to attempt." Youssef elides the term. Youssef ramps up the gendered connotations of the Arabic term "الاسترجال" [*al-istirjāl*] – which means "to act like a man" (Wehr 1994, 329) – by using the word "macho." Hutchins does not. In Arabic, جارية [*jāriyya*], meaning 'slave-girls,' starkly shows how Zeina believes that the Arab women translators are now forced to ally themselves to one of two patriarchal neo-Orientalist extremes: a soldier or a concubine: "انت اما جندي او جارية" [you are either a soldier or a concubine] (ibid). Youssef chooses the word "concubine" (2011, 24) to maintain how Kachachi purposefully situated Zeina thinking in binary and historically situated Islamic/crusader patriarchal/feminised, invader/invaded metaphors. Hutchins uses the more contemporary word "call-girl" (Kachachi 2009b) which overwrites and erases Kachachi's juxtaposition of the Christian Crusades and the US occupation of Iraq as a metaphor, one which is particularly trenchant in the light of Bush's use of "crusade" to describe the US war on terror after the 9/11 attacks. The word 'call-girl' also domesticates the term within quasi-contemporary Anglophone contexts. Is this choice a refusal to link Arab women working for the US army in 2003 to tropes of sexual slavery at times of male-dominated Empire and war? Or did Hutchins simply not 'hear' the critique that Kachachi was getting at? Reading two versions of the same translation certainly helps frame more questions of what we think we are reading – or hearing – when engaging with a translated work when we have no access to the first version. Reading two translations of the same work side by side can also help us map echoes of *different* 'differends' and the possibilities of sensing what feminist translator scholar Kim Wallmach (2006, 4) terms as "what is usually suppressed, namely in the infinity of possibilities, the free play of meanings." I am not suggesting Hutchins or Youssef should have heard otherwise. The differences show different 'hearings' do carry potentially charged political 'differends' as they move in/visibly across languages.

One particularly charged 'differend of hearing,' for instance, relates to how Iraqi dialect, classical Arabic and the act of translation in Iraq move across Zeina's mind as she confronts the realities of war. Zeina entering a

US aircraft is on one such juncture as it represents a point of (political) no return:

"للمرة الاولى في حياتي صعدتُ الطيارة من طيزها هكذا تفتح طائرات الكارغو 17 من الخلف. اما بوزها فكان عريضا مثل كوسج"

[For the first time in my life, I climbed into the plane from up its ass because that's the way the Cargo 17 aircraft opens up, from the back. Its trap, though, gapes widely like a sawfish.]

(Kachachi 2009a, 38)

In Arabic, a reader 'hears' dialect and formal written Arabic colliding through Zeina's chain of thought – a collision between what Zeina says and thinks in dialect alongside the formal Arabic she uses to writes. The dialect term for aircraft "الطيارة" [*al-ṭayyāra*] bumps along with the formal Arabic equivalent for aircraft "طائرات" [*ṭa'irāt*] (ibid) in the same sentence. Juxtaposing the word "بوزها" [*būzhā*], a dialect word for 'mouth,' is the adjective 'wide' "عريضا" ['*aridān*] conjugated using the *tanween*, an accusative construction not often used in this context. The most striking political metaphor is "من طيزها" [*min ṭīzhā*] (ibid): a colloquial Arabic word meaning 'back-end' as well as 'ass' or 'arse.' The formal Arabic 'equivalent' – المؤخرة [*al-mu'akhara*], meaning back-side' or 'bottom' – could have been used by Kachachi in Arabic. But she clearly chose not to. As a quasi-swearword or dialect word used in informal social contexts, طيز [*ṭiz*] features rarely in Arabic-language literary novels. Politically, the word has impact: it communicates the ethical magnitude of Zeina entering Iraq via the US military machine. With a sawfish's mouth evoking an image of a serrated gangway, the translators are shown as willingly walking into the mouth of predatory carnivore, the only exit being getting defecated out through its behind or 'arse.' As well as an allusion to the moral status of the translators collaborating with the US military presence in Iraq, this word evokes very powerful commentary about the US occupation's own point of entry into Iraq under appearances of liberation mired in political expediency. The epigraph which warns against the fresh-looking plant growing from filth reverberates here once more. How do the translators 'hear' this important passage? Hutchins seems to avoid its stark political metaphors.

It was the first time in my life that I had climbed into this type of plane. C-17 cargo planes load from the rear and have a broad snout like a fish's.

(Kachachi 2009b)

Here we read a neutral description of Zeina getting into the aircraft, no mention of her point of entry. By rendering كوسج [sawfish] as "fish," Hutchins removes the predatory metaphor of the military aircraft used by Kachachi in Arabic. Hutchins uses the word "snout," so making the mouth sound more anodyne. Admittedly, a 'sawfish' has a very long mouth and nose which does not resemble the wide 'back' of a Cargo-17. Hutchins presents a much 'cleaner' entry for the translators going into an US army aircraft.

Interestingly, Youssef (Kachachi 2011, 29) engages with the predatory fish metaphor and avoids overt crudeness by using "from behind," a word with more genteel yet also interestingly c/overt political connotations.

> For the first time in my life I boarded a place from its backside. For that's how the C-17 opens, from behind. Its mouth is wide like the jaws of a shark.
>
> (Kachachi 2011, 29)

Her translation of "sawfish" as "like jaws of a shark" echoes the disturbing audible associations of *Jaws*, a series of disaster films about sharks well-known to many US and international readerships. Youssef does not translate the Arabic word طيز [*tiz*] as 'ass' but renders it as "from behind." Her choice is not due to Youssef being unable to attest to swear words in Arabic. Later in the novel, Zeina tells a fellow interpreter offering his photo taken with the US Secretary of Defense to her as a present: "You can put it in your ass" (Kachachi 2009a, 161). As this sentence appears in English in the Arabic version, Youssef supplements this sentence in English translation with "You can stick it up your ass" (Kachachi 2011, 146). Youssef also engages with dialect words of the army interpreters left alone in the army-base in creative ways. She supplements: "هاي شلون ورطة؟ . . . هاي وينهم؟ . . . وين جابونا ونسونا؟" [What sort of mess is this? Where are they? Where have they brought us and forgotten us?] (2009, 44) as "'What kind of mess is this?' 'Where the fuck are they?' 'They just brought us here and forgot about us?'"(2011, 35). Youssef has no aversion, then, to translating crude language into English. Was her choice 'of behind' for طيز [*tiz*] out of concern for US readership sensibilities? Or did she read it as a colloquial word requiring less brashness?

The purpose of reading both translators' interventions so closely is to consider the politics of translation and the ethics of translation for shock-effect in post-2003 Iraq-US literary contexts. Translators engaging with 'shock language' for political effect has been debated at length in feminist translation studies since the early 1990s (Flotow 2009, 2). A frequently cited example is Linda Gaboriau translating the line "Ce soir j'entre dans l'histoire sans relever ma jupe" in the French-language feminist play *Les nefs de sorcières* by Nicole Brossard into English as "Tonight I am entering history without opening my legs" (Flotow 2009, 245). Her decision to ramp up the shock value was supported by Brossard herself but raised many questions on the ethics of feminist translation as activist rewriting. Rosemary Arrojo (1994) asks how far can a (feminist) translator rewrite a text without re-enacting patriarchal writing praxes? Sherry Simon (1996) suggests that that translators can be free in their praxis as long as they are "upfront" about their politics of translation (1996, 36). But in the case of William Hutchins and Nariman Youssef, neither translator overtly articulates their politics of translation in an 'upfront' way. Hutchins has nonetheless the copyright of his English translation – which means he has rights where he posts it. His version's cost-free availability online is more immediately accessible to wider audiences than Youssef's hardcopy version. His interpretation of Zeina's entry into C-17 aircraft, is arguably a politics of aesthetics. His choice to use the phrase "born prostitute," however, could easily be read as a word that Kachachi herself wrote in Arabic. As Kachachi is known for her historical solidarity with women in her literary projects and for not denigrating their choices of survival in Iraq, Hutchins' 'looser' representation of this word emerges as ambivalent and highly risky.

(Re)locating different women's voices in Arabic and English para/translation

At this point, I move on to questions of translator (self-)location, focusing only on Nariman Youssef's engagements not translated by William Hutchins, as the critical receptions of the English version of the novel focus on the 2011 version only. Marcia Lynx-Qualey (2010) comments on the mix of UK and US vocabulary in Youssef's translation. Citing the sentence "I opened [my lunch box] and found a sandwich, a bag of crisps, a Coke and a cookie" (Kachachi 2011, 30), Lynx-Qualey states (2010): "the poor brain falters in the face of an American saying 'crisps' unless he or she is doing a Monty Python impression." That said, Lynx-Qualey very much praises Youssef's other engagements. She lauds "the lovely job Nariman Youssef does with Kachachi's often-lush sentences" and how she creatively works with the many cultural points of reference of the Arabic novel in English translation. Other interesting aspects of the translation not covered by the reviews relate to how Youssef renders the polyphonic aut/her/ship of the novel in English, which is where I start my reading. One example is when Zeina rails against 'the Iraqi woman author,' seemingly a figment of her imagination. She expresses resentment at their relationship being "a duet forced to play on one piano. She [the author] wants us to type together – four hands and twenty fingers" (Kachachi 2011, 26). The following excerpt creates a cleft in any semblance of unity in authorship, implying that the English version is a para/translation of another para/translation: its version in Arabic.

"لكنه تاريخي من قبل ان اولد وانا سليلته وصاحبة الحق فيه مهما بداتُ غريبة وناكرة له
فهل تظن تلك الكاتبة الغشيمة انني ساتخلى لها عن ارثي ,حتى ولو كان وطنية مهلهلة لم تعد
تنفع في شيء . . . ."

[But it [Iraq] is my history before I was born, I am its descendant and I have a right to own it, however strange (foreign) and abhorrent I may appear to it [Iraq]. Does that naive woman writer think I will give over my inheritance to her even if it is worn out nationalism no longer of use for anything?]

(Kachachi 2009a, 28)

It's my history, whether I like it or not. It was mine before I was born. I am its legitimate child, no matter how foreign I may seem. How dare she, that gullible writer, think I'll just hand over my inheritance to her, even if that inheritance is nothing but a tattered piece of nationalism?

(Kachachi 2011, 28)

This train of thought is taken from Zeina's diary, and it is about Zeina reflecting on herself as a product of Iraq's own history. She feels that her own sense of hybridity emerges as at odds with Iraqi identity being configured as 'one' unified thing. Youssef (2011, 28) translates "غريبة وناكرة" [*gharībatān wa nākirtān*] (Kachachi 2009a, 28) as one word, "foreign," leaving out the negative sense of "ناكرة" [*nākira*], which in Arabic means "abhorrent" and "ungrateful" (Wehr 1994, 1171). Youssef also adds "whether I like it or not" (Kachachi 2011, 28) to the phrase "لكنه تاريخي"/ "It's my history," effectively re/writing Zeina's sense of her agency as unwanted but accepted as such. Youssef also shifts the terms of Iraqi "worn out nationalism" by removing "لم تعد تنفع في شيء" (Kachachi 2009a, 28), which translates as 'no longer of use for anything,' a subtle and transformative move which effectively elides nationalism being framed as useless in the Arabic version. Although subtle in premise and political agency, these interventions imply that three rather than two pairs of hands are on Zeina's keyboard –Youssef's along with Zeina's and that the writer who Zeina fears. For an English-language reader, the story reads as told by Zeina or 'the writer' via the 'invisible' translation of Youssef. Youssef's supplementation 'presents' as Zeina's and/or the woman writer's, be she the omniscient author or Kachachi herself. Youssef's c/overt changes are in effect a *mise-en-abime*, a representation of the work within the work (Dällenbach 1989) blurring what (and who) constitutes the inside/outside of this work's telling. Youssef's changes can only be seen when read alongside the Arabic translation. Her c/overt interventions underscore further how Zeina and her writing itself is inherently unstable, a work of her and others' (repeated) self-construction.

In many ways, the instability of the work's self-construction is not surprising in view of what Kachachi has said about the role of history-writing in shaping the receptions of many Iraqis' stories: "history was always written by the authorities. You don't read much about people. And even less about women" (Kachachi cf. Snaije 2014). In the novel, Kachachi often shows how alternative histories can be recorded as spoken and heard outside of 'official' authority sources. One way is by showing how Mosuli Iraqi dialect can reverberate via – and despite the constraints of – formal Arabic, such as the scene when Zeina attends her grandmother's funeral and sees her Mosuli aunts for the first and last time:

"قالت بلهجة موصولية تقلب الراء غيناً
. . . . منو؟ زينة بنت بتول؟ ايمتى جيتي من بغا؟ تعي شمّيتوكي دبسوكي –"

Transliteration: [*Qālat bi lahja mūsūlīa taqlab al-rā' ghaynan* [She said in a Mosuli accent which switches the 'r' to 'gh'] "*Minū? Zīna, bint Betūl? Aymtā jītī mn bagh ā? t'aī shammītūkī dabūskī . . .*"].

(Kachachi 2009, 190, emphasis added)

Their Mosuli accent is transcribed to be 'heard' in formal written Arabic (which I have tried to render through transliteration). The 'r' or الراء sound in Arabic is switched to 'gh' or غ (the *ghayn* sound), as is often used in Mosuli dialect. The spoken feminine is also foregrounded by an addition of the spoken feminine form of address: ". . . . تعي شمّيتوكي دبسوكي" [*t'aī shammītūkī dabūskī*] (2009, 190, emphasis added). While second-person feminine address is often written in less formal settings such as social media messages either in Arabic or a mixture of Latin letters and numerals (Abu Elhija 2014), adding a كي [*kī*] 'looks' and is grammatically incorrect in formal written Arabic. Its para/text is its very presence – a visual sense of dissonance on the printed page. Alongside its 'wrongness' within formal written Arabic, this work reads as 'correct' by how it 'sounds' – as authentic Mosuli dialect spoken between and to women who hear it. Hearing the Mosuli dialect at the funeral is an audible comfort to Zeina. It is an audible 'living' source of memory – Zeina can 'hear' echoes of her dead grandmother's voice when she hears the Mosuli accent spoken by other women. Zeina is in fact transcribing two echoes: of her grandmother and a moment of comfort from the aunt addressing her directly in the Mosuli Iraqi feminine. Writing or rendering these Mosuli expressions otherwise in formal Arabic would only stifle localised feminine echoes, which shows how 'spoken' word is carried while subverting by the authority formal written Arabic. But how to communicate such a textured politics of re/writing in English? Youssef does the following:

> Then in a heavy Mosul accent that rolled the **"r"** she said, "Who? Zeina? Batoul's daughte**rr**? When did you a**rrrrr**ive from ab**rrr**oad? Come he**rrr**e, my dea**rr**, and let me kiss you!"
>
> (2011, 174, emphasis added)

In the absence of feminine address for "you" in English, Youssef uses a slightly generational form of feminised address: "my dea**rr**." The very letter that the Mosuli relatives do *not* pronounce in Arabic, 'r,' conversely becomes the letter foregrounded in English. This makes Mosuli 'seen' as distinct yet 'heard' differently in English. A similar accent visibility technique has also been used by Arabic translator Marilyn Booth (2007) to foreground Arabic accents as visibly and audibly distinct. Far from stereo-typing what Arabic accents sound like in English, Booth carries out this foreignising technique for distinct and at times political effects – first, to avoid flattening the dynamic effect of different Arabic registers colliding with/alongside formal written Arabic on the printed page and, second, to highlight the polyglossia in regions where each speaker marks their location in Arabic and other language/s by localised and socially situated modes of speech. The difference between Booth's and Youssef's technique is that Booth reproduces the same sounds by using English transliterations. In Rajaa Alsanea's novel *Girls of Riyadh* (2005, 17), one protagonist exclaims "شيز سو كورفي!" [*shīz sū kūrfī*], which means what it sounds like: 'she's so curvy.' Booth renders the phrase as "sheez soo kiyirvy" (2007, 204) to highlight the cross-border permutations of English spoken by affluent Saudi women as performances of their diaglossic hybridity and cosmopolitan lifestyle (ibid, 204). Booth's strategy resonates with mimetic approaches to feminist translation, which "focuses on sound and on sound connections, and mimics/renders the sound of the source text, rather than its semantic meaning" (Flotow 2004, 93). Booth's technique works to audibly replicate "Arabenglish" (Booth 2007, 205) to be 'heard' as it is actually spoken in Arabic via English translation. Youssef's strategy differs in that she works to show Mosuli Iraqi Arabic spoken differently but without replicating the exact same sounds in English.

From a political perspective, 'mimetic' feminist translations audibly perform rather than semantically explain the meaning of the source texts as echoes within – and despite – patriarchal language structures. Aesthetically, the purpose is to communicate an experience of language rather than a surface rendering (Flotow 2004, 93). But how does a translator actually do this? As explained by Kolias (1990), the translator can attest to the cadences *s/he hears* by "re-territorialising" (1990, 214) cadences into the text as part of her attestation to her praxis of listening. This means s/he listens to the first text and then to how it could 'sound' in another language. Crucially, the translator shows her/his intervention by attesting to what s/he thinks a target reader is *realistically* able to hear of the sounds from the 'source' text (1990, 217). Kolias' clear distinction between 'source' and 'target' language 'listenings' crucially creates a space where the listening translator renders the source cadences as audible, but as *differently* audible in target language translation. From this perspective, we can read Youssef's decision to switch from 'r' to 'gh.' If Youssef had rendered the passage as: '*Who? Zeina? Batoul's daughtegh? When did you aghive from aghoud? Come heghe my deagh and let me kiss you!*' would not show the cadences of Mosuli dialect colliding against formal Arabic. It rather would show them colliding against English, which was not the political intent of the novel in Arabic. In Arabic, Mosuli dialect sounds resistant, feminine-focused – deviating from and overwriting formal Arabic while interweaving through it in an understandable way. For this point, how Mosuli accent *sounds* in English translation is less significant than its *overt* and *present* relocation in the text. The translation shows Youssef attesting to her role as a Mosuli-English translator mirroring Kachachi's or Zeina's role as a Mosuli-formal Arabic translator in the Arabic version. Kolias states listening translation is "the possibilities that have revealed themselves to the translator in the course of working closely with it" (1990, 217). Youssef's engagement with the novel's politics of (Arabic) language location is working to mirror the novel's disingenuous engagements with the possibility of being mediated by multiple aut/her/ships. In this way, we could read Youssef's intervention as another riff on the c/overt politics of multiple authorships in the Arabic – an audibly c/overt, extra pair of hands playing Zeina's piano.

There is also a third register of language used in the Arabic novel important to consider when reading the English version of the novel – US English. Its presence in the Arabic version raises critical questions on the politics of transferability of US English when moved into the 'English' version of the novel itself. How can the presence of US English as seen and heard in the Arabic version be authentically rendered in English translation? Why, in a novel about Iraq, would the US register in the English version be important, anyway? Precisely because the novel presents the politics of US identity in deliberately provocative ways. Zeina's presence as an American Iraqi is disturbing to everyone, including herself, and is in her view a lost cause: "I couldn't be anything but American. My Iraqiness had abandoned me long ago. . . . I tried to be both but failed" (2011, 163). The Arabic version shows the language of America as overtly 'alien' and foreign to Iraq and enacts its foreignness via the novel's own Arabic script. Phrases such as "Mam, they will take you" (Kachachi 2009a, 58), "Sure go ahead" (2009a, 67) and "Do you speak English" (2009a, 108) rendered in English show the banal and violent interactions of the US army, such as raids on Iraqis' houses (2009, 106), as tersely intrusive and alike but not 'part' of Iraq's story. Zeina writes herself (in Arabic) as exclaiming "Oh my God!" (2009a, 67) and cruder phrases like "Fuck you" (2009a, 130) and "Put it in your ass" (2009a, 161) via English script. She also ventriloquises English words in transliterated Arabic to highlight their and her foreignness to herself and Iraq in Arabic: her reply to a sergeant as "Yes Sir" is written in Arabic script as "يس سير" [*yis sīr*] (2009a, 59). She notices how her Iraqi surname "بهنام" [*bahnām*] is mis/pronounced as "بهنايم" [*bahnāyyam*] by the North American twang of her military colleagues (2009a, 30). In the English version of the novel, another 'English' not at all present at all in the Arabic version finds its way into the English version of the novel which many US readers, such as Lynx-Qualey, found distracting and confusing to the story: UK English. Zeina refers to herself as using a "mobile" (2011, 9) instead of a 'cell phone.' She remembers the local "cleaner" in hometown Detroit (2011, 23) but does not refer to him once as a 'janitor.' The presence of these UK English words is puzzling, as no UK protagonists feature in this novel. The only link to the UK can be found on the "Notes on the Translator" end-page which cites that Youssef "lives and works between Egypt and the UK" (2011, 183). Yet this bio does not explain her choices. We do not know whether the US/UK mix was a decision by Youssef, or perhaps by editor/s. What we end up reading is an evocative novel of an Iraqi-American interpreter whose American voice occasionally gets 'jarred' in translation.

From a mimetic perspective of feminist translation, the presence of the UK words alongside the voice of Zeina – for whatever reason they are there – is a very interesting intervention for many reasons. As the (US) reviewers' reactions attest, it creates a very creative yet possibly unintended political effect. It performs and enacts a barrier preventing (US) readers hearing or reading Zeina as 'fully' American (however 'fully American' is understood to be). Along with the 'alien' presence of US military in Iraqis' lives in post-2003 Iraq and the highly foreignised Mosuli Arabic reverberating both languages, the 'jarring' presence of UK English words underscores that novel has to be read as a *translation* – not as a story about Iraq narrated by a US-Iraqi citizen in transparent (US) English. If Zeina fears 'the Iraq woman writer' and her grandmother are trying to take over her novel, the presence of such 'jarrings' in translation suggest that someone else is afraid that the (US) English language reader may be at risk of taking her story over too. One way of reading the UK register, then, is as an audible trace of resistance to this story of Iraq being overwritten in English as well as in Arabic: for the US to 'go into' Iraq, the US needs interpreters like Zeina. To 'go into' the novel, US readerships need a translator like Youssef. Like Zeina, Youssef translates. Like Zeina, Youssef does not give a seamless translation. The mixing of UK and US English alongside her making visible the Mosuli dialect is a jarring mimetic reminder of the status of translation in both versions of this work. By celebrating these 'jarrings' as productive tensions, not as 'mistakes' on the part of the translator, we can understand why this novel has such varying and specific 'audible' locations which all readerships must read and attend to with care. The presence of more than one English and one Arabic in this novel in effect attests to the politics of this story: an Iraqi woman author whose reference of care first and foremost is Iraq and Iraqis, not necessarily US readerships.

Conclusion: a paradigm of aural feminist translation

Kolias' (1990) praxis of listening translation configures translator choices as attestations which echo or refract different political engagements. By understanding translation as potentially *performing* as well as semantically *explaining* a translation, we can understand the extent to which this novel's 'audible' texture consists of many diverse languages and registers, all ciphered by the act of writing. Mosuli Iraqi dialect is para/translated via formal Arabic, the language of officialdom which would deny its very existence in written form. It is also para/translated through English, a language and script which has few shared references of intelligibility. US military English is para/translated as an unwanted intruder in Arabic script as well as on Iraqi soil. The English version para/translates the presence of US English into its fabric via occasional non-US English to cipher its own status as a translation. It is through these audible textures that the 'differend' of the translator emerges as critical aural and political importance without revealing where the translator 'is.' The ambivalence of translator location is not surprising, as *The American Granddaughter* is a story which troubles all locations of re/writing per se. In 2001, after the 9/11 attacks, US President George W. Bush set out the US-led war on terror on starkly binary terms: "Either you are with us, or you are with the terrorists" (Bush 2001). In the eyes of her Iraq grandmother, Zeina returning to Iraq wearing a US army uniform is also a betrayal – of Iraq's history. Zeina's hopes of helping Iraq as a translator for the US army with the aim of helping Iraq is shown as many things: an act of betrayal and resistance to authority figures, and always politically located. Reading how Nariman Youssef and William Hutchins 'hear' Kachachi's story in Arabic comparatively sheds critical light on how the subtlest of translator 'listenings' can shape or alter how the politics of a novel – and a writer – are read and received new critical locations. Focusing on the audible politics of الحفيدة الأميركية [*The American Granddaughter*] moving across Arabic and English also interrogates why a translator 'listening' without attesting to her/his "movement into alterity" (Godard 1989, 44) carries risks: however radical and creative, many echoes of translator intervention are simply not 'audible' to all readerships. This opens many questions on the aesthetics of intelligibility across languages and their political effects.

Notes

1 "Si je t'oublie, Bagdad." *FranceCulture* www.franceculture.fr/oeuvre-si-je-t-oublie-bagdad-de-inaam-kachachi.html
2 "Si je t'oublie, Bagdad d'Inaam Kachachi." http://missbouquinaix.com/2013/05/06/si-je-toublie-bagdad-dinaam-kachachi-2008
3 " תְּהִלִּים 137 אִם־אֶשְׁכָּחֵךְ יְרוּשָׁלָ͏ִם- תִּשְׁכַּח יְמִינִי: ה: "If I Should Forget You Jerusalem." Psalm 137: 5, (cf. George Phillips. 1846, 552–553)

4   Conversations about 'solidarity among the subaltern'

Betool Khedairi

Iraq has been a land of many peoples, ethnicities and religions for thousands of years, with countless, diverse local cultures in specific regions of Iraq existing for centuries (Salloum 2013). After its rise to power in 1968, the Iraqi Ba'athist government set the parameters of Iraqiness within very specific frameworks of Arabness and Islam in ways which resulted in many radical and tragic changes to long-standing ethnic and sociopolitical make-ups of Iraq, the forced resettlement of Kurdish Iraqis (Aziz 2011) and mass deportation of Iraqis of Shi'a descent being just two examples in the 1970s. The 1970s era, however, is not remembered by all Iraqis as a tragic time: some refer to this era as ايام الخير [*ayyām al-khīr*], 'the days of plenty' – a time of great wealth, increased education and prosperity from oil production revenues. Suad Joseph (1991) recalls during her visits to Iraq in the 1970s, however, that "advantages citizens received from the state were represented less as rights of citizenship, and more as the benefits of loyalty to the head of the party and state" (1991, 178) – in other words, conditional Iraqi belonging and citizenship. This dynamic of belonging applied to Iraqi intellectuals, writers and artists, too, which is why deafeningly absent in most 1970s Iraqi literature is critique of the Iraqi government (Tramontini 2013, 54). This is not to say that Iraqi writers did not express discomfort at the political dynamics underpinning the premise of *khīr* in Iraq. But they did so at great risk. Abdul Sattir Nasir was imprisoned for a year due to publishing his allegorical tale سيدنا الخليفة [*Our Lord, The Caliph*] in 1975 (Moosavi 2015, 16), as this story was read as intimating that a politics of corruption was underpinning the Iraqi Ba'athist government (Mustafa 2008, 101). Iraqi women stories at that time more often than not portrayed Iraqi politics of belonging from the perspectives of class and gender (Ghazoul 2008, 198). Betool Khedairi's novel كم بدت السماء قريبة! [*A Sky So Close!*] (1999; 2001) was the first novel by an Iraqi woman writer which critically portrays ايام الخير [*ayyām al-khīr*] from gendered perspectives of class from a retrospective perspective. In this chapter, I explore how *different* conversations about class during *ayyām al-khīr* interweave this novel across its three languages – formal Arabic, Iraqi dialect and English. I explore why this novel calls on us to reflect on the tensions of its own representation as an Arabic publication and its (post 9/11) English translation, using analytical frameworks of feminist translation that interrogate the dynamics of oppression through language via the configuring of translation as 'a conversation.'

كم بدت السماء قريبة! [*A Sky So Close!*]: critical contexts

Betool Khedairi was born in 1965 and grew up at the time when the Iraqi Ba'athist Party's wider political discourse of pan-Arab nationalism was beginning to take root in full force in Iraq. She also grew up in Iraq during *ayyām al-khīr* as an Iraqi of mixed heritage, her mother Scottish, her father Iraqi. After graduating from the University of Mustansiriya in 1986, Khedairi has lived between Iraq, Jordan and the United Kingdom ever since. Khedairi describes her decision to write stories as a way for her to try to make sense of horrors she could not understand beyond specific political ideologies: "The Iraq in my work is the human Iraq, not only the one of the headlines. . . . My country is bleeding. I will not make negative statements just to prove that I am a liberal intellectual. I search for fairness amidst this chaos" (cf. Voykowitch 2003). Her commitment to representing the lived human element of Iraqi experience in non-categorical ways is one reason why Saadi Simawe (2004) includes Khedairi amongst Iraqi women writers of note in post-2003 Iraq. Her use of humour to show how Iraqis admirably face often impossible choices in everyday life in the most desperate of circumstances is another. Khedairi is also distinguished by the co-collaborative ways that she has published her novels. Neither the English version *A Sky So Close!* (2001) nor كم بدت السماء قريبة! [*A Sky So Close!*] (1999) has an introduction. The Acknowledgements page of the English version, however, shows Khedairi (2001, 243) presenting her novel as the product of many co-collaborations moving across languages, agencies and locations, not just her own work of creativity. She thanks the publishers of the Arabic and the English editions. She expresses gratitude towards her friends and family; her uncle Muhayman Jamil, the translator; her uncle Manar Jamil, proofreader of the translation; and proof-editor Margherita Wilson for "polishing" the text (ibid). Similarly, in the Acknowledgements page of the English version of her second novel غائب (2004), published under the title of as *Absent* (2005), she thanks all involved in both versions' production. In this respect, underpinning Khedairi's writing seems be a recognition of dynamic: a dynamic of languages and books interacting, in conversation with each other rather than 'one' version of a book being 'derivative' to another.

كم بدت السماء قريبة! [*A Sky So Close!*] begins in the 1970s and ends after the 1990–1991 Iraq war. It is told from the perspective of an unnamed protagonist with an Iraqi father and English (not Scottish) mother. The first part begins in the rural village of Al-Zafraniya and ends with the protagonist's friendship with Khaddouja, a girl in the village who dies of bilharzia, a water-borne disease caused by parasitic worms. The second part is about life in Bagh-dad during the 1980–1988 Iran-Iraq war, where the protagonist falls in love and has sex for the first time, and then endures another death of someone close to her and constitutive of her lived sense of identity – her Iraqi father. In the third part, she moves to London with her English mother who is dying of cancer and embarks on a short-lived relationship which results in an abortion, while seeing the 1990 war in Iraq unfold from satellite screens in London cafés. The novel ends with her reading letters from her friends in Baghdad living with international sanctions in Iraq and experiencing what Yasmine Bahrani (2001) describes as "the horror of an Iraq in tatters and a middle class in exile."

Bahraini's focus on representations of class in the novel, and the Iraqi middle class in particular, is an interesting point, particularly because one of the reasons for this novel's critical acclaim in Arabic is its politics of cross-group solidarity, something which Ferial Ghazoul terms as "solidarity among the subaltern" (2008, 198). By this, Ghazoul means that Iraqi women writ-ers' novels often "represent an unconscious affiliation between marginalised groups and non-institutionalised literature" (ibid), Khedairi's novel being one of them. Ghazoul praises, for example, how Khedairi presents the worldview of rural Iraqis in this novel from nuanced perspectives by sensitively showing rural expressions of creativity, such as art, legends and songs inspired by the local environment (ibid). Solidarity-in-alterity is also shown in كم بدت السماء! قريبة [*A Sky So Close!*] by its construction as a novel itself: a *bildungsroman* in which a younger protagonist relates what she 'sees' of a dystopic world, her 'adult self' commenting with the benefit of hindsight. In this respect, this novel can be read as a commentary on the politics of remembering 1970s Iraq as *ayyām al-khīr* – what the protagonist sees of the *khīr* in Iraq at that time shows that this *khīr* was not experienced or enjoyed by all Iraqis alike. At the same time, not one political reference can be found in this story about rural Iraq, which opens the question of how its sharp critique can be 'sensed-as-read' and how it shifts and moves across languages.

Reading Iraqi women writing 'solidarity among the subaltern' across languages

One key starting point for this question is looking at what representations of 'solidarity among the subaltern' in Iraqi women's stories, including Khedairi's, mean in Iraqi literary contexts. Ferial Ghazoul's turn of phrase seems slightly odd when we consider that many Iraqi women writers situate their stories – through writing – from the perspectives of women, some of whom do not write. This issue of asymmetrical dynamics of representation between different identity groups (or constituencies) has come up time and time again in post/colonial fields of inquiry, notably subaltern studies. The seminal article 'Can the Subaltern Speak?' (1988) by Gayatri Spivak suggests that any act of writing about 'subaltern' identities is likely to reiterate dynamics of power which configure groups as 'subaltern' in the first place, which seems to preclude that a writer can never fully write 'as' one. Emmanuel Levinas' conceptualisation of 'alterity' (in French, 'l'alterité') helps us consider what writing 'solidarity among the subaltern' could mean in Iraqi women's literature, with Khedairi's novel كم بدت السماء قريبة! [*A Sky So Close!*] as a point of focus. According to Levinas (1987), 'alterity' is a sense of otherness arising when a person becomes aware of 'someone' or 'something' else in the world. A person becomes conscious of her own sense of 'self' as 'outside' of herself, *as part of* her/self while outside of her/self (1987, 27). In Iraqi women's literature, this process of feeling 'other' is often shown by representations of a protagonist as a child, experiencing herself in relation to the world for the first time and relating what she experiences from different focalised perspectives. So as well as one child mediating herself as 'I,' children in Iraqi women's literature are often 'presented' as 'representing' perspectives of other protagonists. Levinas goes further on to explain that a person's conscious *awareness* of her/his *own* 'alterity' as an experience mediated by others comes about when s/he 'sees' others as 'other' to herself *while knowing* s/he is also 'being seen' by 'others' as 'other' to them in similar ways (1987, 50). In Iraqi women's literature this awareness of alterity is often manifested by the child relaying how s/he understands or doesn't understand about her life in Iraq for various reasons, such as lack of knowledge (due to being a child), speaking a different language, knowing that s/he herself may be 'strange' or 'other' to others for different reasons.

This notion of alterity is what helps us understand why representations of the rural background in كم بدت السماء قريبة! [*A Sky So Close!*] through different language registers is a crucial aspect of this novel's representations of cross-group solidarities and its political critique. The main protagonist represents the spoken words of rural Iraqis using formal written Arabic, a language which forecloses their 'spoken' Arabic as 'wrong' but not unintelligible within it. She herself experiences marginality and alterity in three different languages – formal Arabic, spoken Iraqi dialect and English – none of which she seems to 'master.' She later realises that many other people in Iraq do not 'fully' master all of these languages due to their own different (political) experiences of alterity – ethnic background, social class and essentialised notions of Iraqi identity. Some 'languages,' such as written Arabic and English, for example, configure some 'groups' in Iraq as mobile (middle class). Speaking in Iraqi dialect and no other language signals, however, that a person is rural and 'expected' (by the middle classes) to remain illiterate, poor and tied to the land, as were their ancestors for centuries. For Levinas (1987), a conscious awareness of one's own alterity – whatever this may be – helps people nonetheless feel a connective sense of ethical separation but also ethical linkage between 'I' and 'other.' This is why Levinas argues that one person 'turns' towards 'another' even when experiences of alterity may differ greatly between them and may not be mutually intelligible (1987, 50). The conscious choice by Iraqi writers to 'turn' to other marginalised Iraqis could thus be understood in such contexts. This conceptualisation of alterity or الغيرية [al-ghayriyya] (the word for 'alterity' in Arabic) also helps us understand why writing a story such as [كم بدت السماء قريبة!] [*A Sky So Close!*] can be configured as a radical act of political intervention at its time as it relates to one of the most charged areas of Iraqi representation – the role of the middle classes at different points in Iraq's history and the politics of Iraqis in different socio-political locations writing stories as a result. In Khedairi's novel, written language – along with the *beaux-arts* of art, colours and dance – is presented as transformative but exclusive to its middle-class protagonists and a means to move on to other places. Rural Iraqis are shown in this novel as having their own local language and aesthetic traditions but little space for social mobility. They just 'stay where they are.' And as the novel shows, the consequences of having less access to mobility, knowledge and education prove tragically lethal, a point I return to later.

So why read this novel's representations of rural alterity in Arabic and English using intersectional frameworks of feminist translation, and why focus on the first section about 1970s rural Iraq? As discussed previously, 'man-made' languages – and the realities they shape – oblige women and other marginalised identities to discursively conceptualise themselves within the hegemonic patriarchal premises that shape them as 'other' or in situations of 'alterity' (Godard 1989; De Lotbinière-Harwood 1991). For many feminist translators, awareness of this challenge presents an opportunity for women – and translators – to 'turn towards' language to expose the unseen hegemonies shaping lived realities, including the hegemonies of translation itself (Castro & Ergun 2018). In practical terms, this can mean adopting "a *counter*-practice of translation . . . working in the gaps and silences of translation and underscoring unequal relations" in ways which disrupt "hegemonic narratives about gender, feminism and the subaltern" (Costa 2014, 135). One very obvious example of disrupting hegemonies of translation is to create pathways to interrogate how and why some neo/colonial languages dominate the global publishing translation market notably as 'transit' languages between one language, such as between Arabic and another language (Loucif 2012). But what could translation configured as 'a conversation about' hegemonic narratives about gender and other alterities enact as an analytical praxis? According to María Reimóndez (2017), it involves consciously working with and questioning what and who is reified – and elided – by languages of translation. It involves questioning how 'alterity' is re/configured and represented across languages and mediums of representation and why (2017, 44). Reimóndez cites Tamil and Galician feminist activists co-compiling a book and a film in (and across) Tamil and Galician with the (hegemonic) medium of English between them a shared point of interrogation (2017, 52) as an example of this praxis. So far, however such an approach or conversation about hegemonic language and translation is yet to be considered alongside Iraqi women's literature, and perhaps كم بدت! السماء قريبة [*A Sky So Close!*] may be a good place to start. Part of embarking on such a conversation, however, is to acknowledge the charged politics of writing itself in relation to this novel, an issue that Khedairi seems to have been aware of but not able to resolve herself:

> I tend to offer a collection of photos or sketches to the reader. I lay them on the table and I leave. . . . Every reader has the opportunity to change the photos as he sees them. I don't like to impose my thoughts on someone.
>
> (cf. Leinwand 2003)

Khedairi's framing of her own writing (in English) as 'photos or sketches,' then 'leaving,' suggests that she is trying to communicate her sense of ambivalence towards all written modes of representation. She consigns some of the story's meaning-making to the reader as a co-collaborative participant in her conversations about Iraq – in other words, she recognises that what people will understand or 'take' from her novel will depend on their 'epitexts of understanding.' Her awareness of how words can operate as instruments of power relates well to كم بدت السماء قريبة! [*A Sky So Close!*]: it is a story about the alterity of Iraqis told from a perspective of a middle-class Iraqi-English girl who understands (and often misunderstands) three languages of power and mediation – formal written Arabic, spoken Iraqi dialect and English. The dynamics of how experiences of social class intersect all of three of these languages is a recurrent theme in this novel. One example is when the protagonist introduces her best friend, Khaddouja, explaining why she will go to school, and why Khaddouja never will:

> –"من هي خدوجة؟
> هي في المزرعة ولا تذهب الى المدرسة , لأنها حافية.
> صدّقتُ حينها أن من لا يرتدي حذاءً لا يذهب الى المدرسة"
>
> (1999, 9)

And in English:

> – Who's Khaddouja?
> – She lives near our farmhouse. She doesn't go to school because she has no shoes.
> I believed then that children who didn't have shoes didn't go to school.
>
> (2001, 6)

In the Arabic version, the protagonist names Khaddouja as حافية [bare-foot]. This is an essentialising term of rural identity communicated to her by adults that configures Khaddouja and other 'barefoot' children like her as never destined for education – even if her family had had the means to buy shoes. In English, the reason is 'explained', the essentialising Arabic term gently elided. In both versions of this excerpt, the older version of the protagonist seems to chide her younger self for her naivety.

This scene is an important one because it prefigures and explains why the novel – and its cross-constituency representations – are to be presented solely from the perspective of the protagonist. She is the one who knows all three languages – formal Arabic, spoken Iraqi dialect and English – but as a child, not perfectly. Khaddouja however cannot 'write' this story at all – as she cannot go to school to learn write. Reading this scene highlights the very tension of Khedairi configuring her stories as 'photos' as images liable to be changed at will. To 'see' or 'change' her photos in the first place, someone must be able to 'read' them – an action fore/closed to 1970s rural Iraqis. Part of a conversation then is to consider what it may mean to 'change' the photos of Iraqi rurality as they move across languages, and who has the 'right' and 'ability' to change them, in the context of this story. As gender is not the only field of inquiry when analysing – and interrogating – the power dynamics influencing how different works, discourses and literary traditions move across languages, I focus on a selection of Khedairi's cross-class engagements between the protagonist and her best friend Khaddouja in Arabic and English translation and ask why these engagements open up different conversations about the paradigm of 'reading' these scenes as photos of 1970s Iraq. Amongst other questions I ask: How and why do scenes or 'photos' of Iraqi rurality shift via translation? What are the politics effects of such para/translation? Why do some the 'photos' of Iraqi alterity appear more 'difficult' for some readerships to 'read' or 'see' clearly than others.

Interactions as conversations: para/translating Iraqi rurality

From a perspective of feminist para/translation, I 'read' the very first 'words' (or 'photos') of this novel – the novel's title كم بدت السماء قريبة! – 'how close the sky seems,' translated into English as *A Sky So Close*. The title refers to an early scene remembered by the protagonist of playing on an old swing in the village of Al-Zafraniya. After Khaddouja plays on the swing, she does, too:

" جاء دوري. ركلتُ الهواء بقدميّ . . . ارتفعتُ إلى أعلى . . . ركلتُ أقوى . . . ارتفعت
. . . أعلى . . . سبحتُ في فضاء . . . أطرتني زرقة حليبية.كل النخيل تحت قدميّ الحافيتين
كم الشمس تسبح في مياه النهر . . (. .) . . أرتفع . . . . استنشقت خط الأفق . . . عندها . . .
بدت السماء قريبة!!"

[(Then) it was my turn. I kicked the air with my feet. . . . I went up higher . . . I kicked harder . . . I went up higher . . . and was swimming in space . . . framed by a milky blue. All the palm trees are beneath my bare feet . . . the sun is floating, swimming in the river. I go higher. I breathe in the horizon. . . . At the horizon . . . how close the sky seemed!]
(1999, 17)

As soon as this excerpt is 'seen' (as read), the connection between the title and the story is made. Brimming with imaginative, colour-based metaphor, the final exclamation: !كم بدت السماء قريبة [A Sky So Close!] is a spontaneous moment of wonder, the protagonist's joy of swinging free in the air. The protagonist noticing her own bare feet and the landscape she sees as she swings (تحت قدميّ الحافيتين) is a subtle reference to and subversion of 'bare-footed-ness' as an identity marker that configures anyone to a life of poverty with no schooling, with no space to imagine the world differently. The novel's English title – *A Sky So Close* – similarly anchors its title to the exact same scene:

I rise higher toward the heavens . . . I breathe in the horizon . . . then. . . . A sky so close!

(2001, 16)

What the title doesn't tell us is that the scene does not end there. As the protagonist swings higher and higher, the rope snaps, catapulting her to the ground – bringing her back to earth with a bump (1999, 18). Undeterred, she and Khaddouja lose themselves in another game with the local village children (1999, 19). The scene concludes with a nostalgic frame of memory of her times with Khaddouja:

"هكذا، كانت أيامي معها, سلسلة من أيام جمعة لا تتشابه"

[That was how my days were with her (Khaddouja): a series of Fridays, one day never like the other.]
(1999, 19)

After reading further memories of the protagonist of her times with Khaddouja, we can appreciate why Khedairi infuses 'reading' and 'writing' with potential ambivalence. It is Khaddouja who brokers kinetic knowledge to her 'outsider' friend by teaching her local words and songs. Khaddouja introduces her (and the reader) to village life: how her mother makes bread, how the elderly grandmother, Bibi Hijjia, tells local legends and local folk songs to all the children, and tells them where they should play. Through getting to know village life, Iraqi dialect words used in rural spaces are relayed to the protagonist, and by proxy to the reader of the story: "سميط" / "sameet" (1999, 24; 2001, 24) or sesame-skin dough rings, "الجب" / "hib" (1999, 34; 2001, 37), a water jug and "شكيك" / "chickeek" (1999, 33; 2001, 35), a local river cactus, to name a few. The subtle shifts in these words' translation allude to the liminal presence of 'a reader' reading in different languages. Localised Iraqi dialect words such as "زلنطح" / "snails" (1999, 14; 2001, 11) and a local herb called "تشيخ صملّه" / "sheikh smalleh" (1999, 19; 2001, 16) are explained in both versions. Less localised Arabic words such as "الفوطة السوداء" [black veil] (1999, 24) are not explained in Arabic but are explained in the English version, as in "futa – the black veil covering her hair" (2001, 25), para/translating expectations of varying limits of cultural understanding. In response, the protagonist listens and commits this knowledge to memory across two languages. Khaddouja's words 'in writing' ensure that important details of rural cultures in Iraq are recorded for 'others' to read (or 'see'), whether they are read in Arabic or English para/translation. The protagonist uses the power of her imagination, nourished by her knowledge of poetic metaphors, to evoke some of the most memorable 'photos' of rural Iraq that we find in this novel. She describes Bibi Hijjia, the eldest person in the village, for example: "وجهها صورة مكثفة للشقوق المتفطرة لذلك الجدار و ما بيبي الحجية إلا امتداذ له" (1999, 24), "the overcrowded wrinkles on her face resemble the jagged fissures in the mud wall behind her. As a condensed image of the cracks, Bibi Hijjia appears to be merely an extension of the wall" (2001, 25). This description evokes many questions from intersectional perspective of feminist translation: what sort of 'photo' is being offered to a reader here in both languages? Is it a celebration that Bibi Hijjia is connected to the landscape? A metaphor of her (dark) skin colour? The protagonist is 'taking' knowledge from Khaddouja and her family to then represent it to be 'seen' by others. Bibi Hijjia could not comment on this photo because the story configures rural Iraqis as never able to read, so not able to 'see' it.

This image of Bibi Hijjia 'drawn by words' also calls to mind conversations of the politics of the visual in relation to gendered representations of power. As astutely observed by Peggy Phelan (1993): "if representational visibility equals power, then almost-naked young white women would be running Western culture" (1993, 10). From a more intersectional perspective, Sharbat Gula, known as 'the Afghani girl with the green eyes' featured on the June 1985 issue of *National Geographic*, also highlights the lack of reciprocity within dynamics of power which privilege the visual. In theory, Gula should have had the space to choose where she lived as a refugee after becoming "the unwitting posterchild for the plight of thousands of Afghan refugees streaming into Pakistan" (Strochlic 2017). Instead, she was arrested and detained in Pakistan in 2016, the place she had lived most her life and told to leave for Afghanistan. She was accorded a luxury home in Afghanistan. But this was an option 'granted' to her and not of her own choice. In the novel [*A* كم بدت السماء قريبة !*Sky So Close!*], the protagonist's memories that initially create a lush, detailed panorama of rural Iraq seem to celebrate and 'preserve' histories of 1970s Iraq. Yet these 'photos' do not show how rural Iraqis may have been questioning the parameters of their life, and like Gula, making their own choices of where to live. And many rural Iraqis did question the parameters of rural life – a questioning, made apparent through rural migration flows into the cities in Iraq from the 1960s, the turbulent politics of which are described in Samira Al-Mana's novel القامعون [*The Oppressors*] (1968/1997) and which were an important factor in the Iraqi Ba'athist Party's rise to power (Davis 2005). Admittedly, the protagonist makes efforts to help Khaddouja read and write (1999, 51; 2001, 45) and offers to lend her shoes to help her go to school. This does not change Khaddouja's lived reality in any way: as a rural Iraqi, she must help her mother by working on the land and staying where she is. Khaddouja's help to the protagonist has much more practical effects. She helps the protagonist – and the reader in Arabic and English – better understand life in 1970s Iraq.

Gayatri Spivak (2000) terms such ambivalences of reciprocities in (written) representation and enacting a politics of solidarity (in contexts of her own rural activism) as a "haunting" (2000, 104): a haunting of "the impossibility of fully realising the ethical" (ibid, 105) when one collaborator's needs are clearly being met more than that of the other collaborator (ibid). Specific examples of how this tension manifests itself via writing and translation are made apparent in Khedairi's novel, too, as shown in this example, which shows the protagonist and Khaddouja in conversation with each other. At this point in the novel, the young protagonist is going to school, and Khaddouja has to work on the land:

"لماذا تحمل بقرتكم في اسفل بطنها كيسا منفوخا تتدلى منه اصابع كثيرة؟
– حتى يكفينا الحليب ولا نجوع.
– امس رأيت **الكتاكيت** تشرب من اصابع البقرة
اجابتني بحدة دون تردد:
– لا تكذبين! الفرخ ما **ينوش ديس البقرة**!
وَضَعَتْ طفلتي البرّية حدا لخيالاتي من رسوم متحركة كنت اتبادلها مع أمي"
(1999, 50)

"Why does your cow have a swollen bag underneath her stomach with so many fingers hanging down from it?"

"So that we have enough milk and don't go hungry."

"Yesterday, I saw the chicks drinking from the cow's fingers."

She retorted without hesitation:

"Don't lie! Chicks can't reach the cow's udders!"

So my wild childhood friend put an end to my visions of moving images which I exchanged with my mother.

(2001, 55)

Here dynamics of cultural frames of reference enact the tension, and they also seem to relate to language. Khaddouja configures the cow as a means for survival. The protagonist is simply trying to imagine life differently, inspired by the stories of her imagination that she talks of with her mother at night. Khaddouja's first response is matter of fact. She does not correct her friend's choice of word "اصابع" [fingers] to describe the cow's udders. She explains calmly that the family needs the cow in order to live. After hearing her friend's flight of imagination about the chicks drinking milk from the "fingers," Khaddouja's reply is much more impatient. She uses Iraqi dialect words – "ينوش" [reach up] and "ديس البقرة" [the cow's udders], correcting' the word"اصابع" [fingers] and relegating the protagonist's flights of imagination to the realm of absurdity – the realm of the real is where the cow is a lifeline. The protagonist does not reply – she seems instead to reply to 'a reader.' Her train of thought refers to Khaddouja as"طفلتي البرية" [my wild girl-child] (1999, 50) which is slightly odd – why would the protagonist frame Khaddouja as 'less adult' than herself? In English, this odd turn of phrase shifts to "my wild childhood friend" (2001, 55). In any case, the presence of the word 'wild' in both versions reveals intersecting class dynamics arising between the two girls. Spivak terms such dynamics as an 'aporia' (Spivak 2000): the tense lived reality of each girl precluding her ability to fully grasp the reality lived by the other. They are literally living and imagining different worlds. It is important to note that another conversation relating to 'alterity' is also emergent here – both girls are still 'turning towards' each other. For Khaddouja, however her sense of reality is the harsh realities of facing rural survival. For the protagonist, her reality is slowly shifting via her educated and literate childhood imagination.

The politics of shifts of imagination moving across registers and class is not so noticeable in the English version, as Khaddouja's and the protagonist's registers do not read as 'differing' from each other in the same way that they do in Arabic. English as the other (hegemonic) cipher of the protagonist's imagination does, however, bring out the tension in other ways, when the protagonist tells the first joke that she has ever heard in English to Khaddouja but in Arabic. It is a well-known joke about a mother tomato crossing the road with a baby tomato, and it ends with the punchline "Come on, ketchup!" (catch up). In Arabic, the protagonist changes the "ketchup" punchline to "باي باي معجون" [Bye-Bye Paste] (2001, 50), which is why Khaddouja does not 'get' it. The pun is lost by the protagonist's clunky English-Arabic translation, and it is not funny at all. In the English version, the 'real' version, "Come on, ketchup!" (2001, 56) is used instead but Khaddouja still does not get the joke. In both versions, the protagonist is devastated that Khaddouja does not get the joke and takes it as a foreboding of bad things to come:

"لم تضحك! عندها فقط، أدركتُ ان بعض الأشياء بدأت تتغير"
(1999, 50)

She didn't laugh! It was only then that I realised that things had started to change.

(2001, 56)

Reading two versions of the novel as 'in conversation' with each other shows how fissures of alterity between the two girls emerge differently due to expressions of alterity moving across languages themselves differently. In the Arabic version, the impossibilities of the joke's representation show how some ways of thinking, such as understandings of puns, 'cannot' translate. The two girls think in two different languages, the protagonist at times trying to mix the two. In the Arabic version, we see that the protagonist's fear relates to her own sense of 'alterity' in rural Iraq. As 'the foreign woman's daughter' (as the villagers call her), she fears that her friendship with Khaddouja and her connection with the village may be changing. In the English version, it is, however, Khaddouja's alterity which is more foregrounded. She simply does not 'get' the joke. Khaddouja's reaction is understandable in 1970s rural Iraq: 'ketchup' as a branded (US) condiment may have been unfamiliar to her. Getting a joke also needs a leap of imagination – here, tomatoes talking and crossing the road – a leap that Khaddouja may not consider, as shown in her reaction to her friend's imaginings about the cow and chicks. The protagonist's sense of foreboding can be read as a conversation about something which goes much deeper than language itself. One girl 'telling' a joke and another girl not being able to 'get' it show how thousands of Iraqis clearly inhabited the same space during أيام الخير [*ayyām al-khīr*] in Iraq – but for reasons of class, poverty and privilege, each 'class' of Iraq and their children live this space very differently.

In the examples I discussed so far, spoken Iraqi dialect and English seem to have been accommodated to varying degrees by formal written Arabic. As the realities of each girl's lived experience of poverty and privilege become more apparent, formal Arabic takes a less benevolent turn. The power – and sinister nature – of formal Arabic as deployed in the novel becomes crystal clear when Khaddouja has suddenly died from bilharzia. The protagonist is devastated. She cannot believe that Khaddouja, her life-long friend, is dead and cannot imagine her life without her. Yet instead of using their formal 'educated' Arabic language to share and bring comfort to sadness, the (educated) adults around her use their adult formal Arabic to 'explain away' Khaddouja and to convince her that Khaddouja's death can be 'rationalised':

"الموت وخدّوجة . . . أخفقتُ في الربط بينهما!
حاولوا إقناعي ان مرض البلهارزيا قتلها مثلما يفعل عادة بالاطفال في تلك الانحاء. شرحوا
لي أنها كيف تبولت دماء كثيرة في مياه السواقي مما أودى بحياتها. امي سارعت الى
عرضي على الطبيب."

[Death and Khaddouja . . . I failed to make the link between the two! They tried to convince me that bilharzia had killed her, just as it usually does children in those areas. They explained how her urinating so much blood in the irrigation waters had led to her death. My mother rushed to get me checked over by the doctor.]

(1999, 59–60)

And it is at this point in the novel that the full (political) weight of Khedairi's commentary on her story-writing being 'photos and sketches' is revealed, and why some people can 'see' these photos of 1970s Iraq and why some perhaps 'cannot.' This photo is constructed to be shown to a very specific set of spectators – those who can read and in order to 'see' it. In this 'photo,' the languages and registers of 'the educated' are subtly being called to account – and anyone to daring to change what this 'picture' was showing here may have to accept that they are in denial or perhaps don't care about the fate of this particular girl, Khaddouja. Particularly shocking here is how the adults believe that framing Khaddouja's death as an everyday statistical banality in formal Arabic would make her death easier for the protagonist as another child to 'understand': "مثلما يفعل عادة بالاطفال في تلك الانحاء" [just as it usually does to children in these areas]. The power of formal Arabic to mediate the injustices of the machinations of (political) power was always apparent at the start of the novel but not as clear or 'readable' to her as a young child at the time: why Khaddouja did not go to school because she was حافية [bare-foot] (1999, 9; 2001, 6); why Khaddouja 'had' to stay on the land and work; why she did not have the luxury of literary imagination, that is, to imagine tomatoes talking and walking across the road. At the point of Khaddouja's death however, formal Arabic is exposed as an instrument by which the privilege of some and the suffering of many was 'explained away' something of why *ayyām al-khīr* in 1970s Iraq were not *khīr* for all.

In the English version, the adults 'rationalise' Khaddouja's death in a similar way, with some other subtle shifts of reference:

Death and Khaddouja . . . I was unable to link the two! They told me that she'd come down with bilharziasis. They explained to me that her lifeblood had drained away every time she urinated into the irrigation ditches and that the continuing blood loss had ended her life. My mother rushed me to the doctor even though you had told her in a condescending tone that it was not possible for me to acquire the disease through human contact. The only way I could catch Khaddouja's disease was by doing what she had done – wading in stagnant water.

(2001, 66)

Here Khaddouja's death shifts from generalisations about (rural) statistics to the *personal* actions of individuated people in the English version. It is *Khaddouja* who waded in stagnant water and paid the price of ignorance. The father patronising the mother para/translates other alterities: he cannot imagine that his own daughter would wade in stagnant waters. As the protagonist's mother knew that her daughter – against her wishes – played with Khaddouja bare-foot on the land, her fear for her daughter's health is not misplaced. For her father as a middle-class father working away from home Khaddouja's death can be 'explained away' by a rural ignorance he cannot pertain to his own daughter. As explained by Hannah Arendt (1963/2005) and Philip Zimbardo (2007), languages of authority – and those using them – often frame instances of injustice within similarly banal or rationalising frames of explanation. One way is to (counter)read such 'languages' is by exposing the very systemic operations of power that allow such dehumanising practices of power to flourish, via their very banality. While both Arendt and Zimbardo were referring to representations of genocide and terror in totalitarian regimes, their commentary is very useful to understand the politics of representation in Khedairi's novel here. This form of formal Arabic is 'seen' – to be 'read' – as the tool or cipher by which 'unavoidable' conditions – poverty, lack of education and poor health – *were often shown as* explained away in 1970s Iraq and ignored by many. During the 1980–1988 Iran-Iraq war, however, many more deaths and disappearances will be shown in the second part of the novel as similarly 'explained away' via interminable TV broadcasts about Iraq's many victories. This scene – and the horrifically banal frameworks of its linguistic presentation – is a powerfully subtle commentary on *ayyām al-khīr* as a nostalgic frame of reference in both languages. It prefigures the more all-encompassing politics of war which the novel later shows would pervade Iraq further decades later – the *khir* or 'prosperity' in Iraq running out, and never equitably shared in Iraqi society anyway.

From an intersectional perspective of feminist translation which works to go beyond the notion of 'written' language itself, we can read this scene as a searing 'photo' or 'sketch' of how the politics of 'banality' towards rural Iraqis may have 'worked' within Iraqi society at that time – through languages. It addresses 'who' may have been part of why banality 'worked.'

Nadje Al-Ali explains the role played the Iraqi middle classes during *ayyām al-khīr* in much more explicitly critical terms:

> For the expanding middle class ... in terms of social-economic rights ... access to education, health care, having a house, a freezer, a car, people could do quite well if they didn't open up their mouths. This was all in the 1970s.
>
> (cf. Wing 2013)

While Al-Ali's commentary is salient, it still needs to be couched alongside the recognition of what usually did happened to people in Iraq who did 'open up their mouths' in the 1970s. They tended to 'disappear' (Mushatat 1986; Moosavi 2015). The framing of Khaddouja's death within the mimetic clichés of hegemonic power thus co-create some space for reflection in both versions of the novel: why was Khaddouja's death explained away by the middle-class Iraqis living in the same village this way? Was it because they did not care? Did they believe that citing statistics could mitigate the tragedy of her death? Or was it that her death in that public sphere could not be mediated otherwise in Iraq? Read in this light, 'authority-speak' mimetically ciphers how fear may have played its part in the systematic abuse and marginalisation of different *constituencies* of Iraqis. The overt blatancy of formal Arabic's own deperson-alising flawlessness along with the wider politics of fear in Iraq testifies to its own injustice. In this passage, the protagonist is hearing the banal, non-critical terms of formal Arabic reference used by the adults and does not comment on them directly. Her reply is wordless – deep depression, something which can only be 'dealt with' by moving away from the village. Again, it is middle-class affluence which allows a possibility of flight from memory. As for the people in the village, Khaddouja's family, nothing more in this novel is said.

Cross-language representations of Iraqi alterity and solidarity

The dynamic of the novel's political critiques of *ayyām al-khīr* is an important point of discussion for this novel but has not been attended to in detail in recent scholarly contexts. The English version, *A Sky So Close* (2011), for the most part has been read as a 'war novel' with good reason (Masmoudi 2010; 2015) – the second and third part of the novel are dominated by the impact of the 1980–1988 Iran-Iraq war and the 1990–1991 war in the protagonist and her family. The early reviews of the English version seem to connect the story to the wider politics of the United States after the 9/11 attacks, even though the Arabic version was published in 1999, two years earlier, and its 2001 English version would have long been in process before its actual date of publication in 2001. Elizabeth

Robert-Zibbels (2002) praised the novel: "With Iraq being George Bush's next target in his War on Terrorism, *A Sky So Close* is an interesting and timely look at the life of one woman in a country oppressed by the ideologies of hegemonic nations." Sonya Knox (2003) – a Beirut-based reviewer – however, complained that the novel did not go far enough in terms of its political critique:

> By avoiding any mention of Iraq's domestic politics but repeating in detail both reports from the Iran-Iraq war and the 1991 Gulf War and the effects of the sanctions on the Iraqi people, Khedairi mitigates the sense of 'telling the truth' that the novel's emotional depth depends on.

Knox also claims that the translation is stilted, which intimates the novel's political critique of formal written Arabic was not always easily readable in English. Each reviewer also seems to read the novel alongside a yardstick of what s/he believes that Iraqi fiction in English translation should, in her/his view, 'do.' As observed by Arta Khakpour, Mohammad Khorrami and Shouleh Vatanbadi in *Moments of Silence* (2016), a collection that focuses on aesthetic and literary (self-)representations in the 1980–1988 Iran-Iraq war, "the facts of war, even when agreed upon, do not translate to simple truths" (2016, 3). It seems that some elements of this novel did not 'translate' simply either. After all, no one in 1999 could have predicted the events of 9/11 in the United States. This raises questions on how paradigms of intersectional feminist translation analysis as a conversation can 'work' when some political premises and later readings of a conversation cannot be mutually known.

With the second and third part of the novel focusing on war in Iraq, approaching translation as 'a conversation' can nonetheless help us see why the title of the novel – !كم بدت السماء قريبة [*A Sky So Close!*] – is one of the most key points of interaction in this novel. The title preserves the memory of Khaddouja and playing on the swing. It opens a conversation, in my view, on the radical power of 'the house of fiction' to subvert the 'statistic-speak' which rationalises her death as one amongst many rural deaths. Khaddouja's echoing presence in the title calls a reader in both languages to 'see' these words (photos) and ask some questions that cannot be answered but which still need to be asked: why is a story about 1970s rural Iraq being told by someone with much less knowledge of rural Iraqi life than Khaddouja? What happened to the others in the village left behind after the protagonist's move to Baghdad? What other stories of *ayyām al-khīr* in Iraq have disappeared due to lack of representation across different languages and media? First written in 1999, this is no '9/11' or 'post-2003' Iraqi novel. It is a polyphonic story which works to subtly 'interrogate' how one Iraqi woman writer has tried to 'write' the politics of unrepresented alterities in Iraq alongside the politics of why it was impossible for her to do so. Reading the fault lines of the novel's representations as part of its para/translatory meaning-making help us open a conversation about why the reading class of Iraqis are connectively implicated in the novel in Arabic and English. Her representations of Iraqi alterity moving across written languages 'show' or 'sketch' Khedairi, in my view, as painfully aware that she is implicating herself, too. If we configure Khedairi's representations of rural Iraqi experience as "a *counter*-practice of translation" (Costa 2014, 135), we can appreciate this novel as c/overtly presenting – via formal Arabic and English – para/translations of an Iraqi rural imagination as traces captured in polyphonic re/writing.

While I have not addressed the timing of this novel's publication in 2001 in the United States, the issue of representing Iraqi alterities in English translation alongside the prevalence of US media influence still remains one unanswered question in my analysis of the first section of this novel. As stated earlier, ‏كم بدت السماء قريبة!‏ [*A Sky So Close!*] has been critically acclaimed as a 'war novel' in such contexts in English. It is certainly that and more: the story is a trenchant critique of interlocking oppressions of state becoming entrenched in the Iraqi public sphere and important example of story-telling, where memories, real or imagined, play their part in ensuring the most minute 'photos' of rural life in Iraq are documented. The subtle shifts in translation suggest that the English version is calling to audiences to 'turn to' or remember Iraq from different perspectives to its Arabic version. Claire Beckett's (parodic) photo exhibition *Simulating Iraq* reminds that this call would have presented some challenges. Beckett's exhibition displayed photos of US citizens playing the roles 'Iraqi people' which were commonplace in the (post-9/11 and post-2003) US media: Iraqis in the street, Iraqi women (played by American men) sitting drinking tea together. (Beckett & Banai 2012). That US citizens were 'playing the role' of Iraqis as presumable 'imagined' by many in the (post-2003) US raises an important point about the extent that Khedairi's representations of Iraqi alterity in 1970s rural Iraq could ever have been read as 'intelligible' through the lens of the charged contexts of post-9/11 US political involvement in Iraq. Part of the political point of Beckett's work was to highlight how little about specific constituencies of Iraqis *could* be known in post-9/11 and post-2003 America. Her exhibition also raises thought-provoking questions on the politics of publishing Iraqi women's literature in English translation at specific points in time. Which localised politics of Iraqi women's stories possible to read in English translation? Are the lines of confrontation enacted impossible to negotiate? Khedairi's novel calls on us to read the shifting representations of differently marginalised groups 'living' *ayyām al-khīr* in 1970s Iraq across Arabic and English as a vital part of such questions.

5 Re/writing confrontations in translation

Alia Mamdouh and Hadiya Hussein

Many Iraqi women writers represent a distinct politics of commitment to all constituencies of Iraqi – diaspora exile, 'informant,' rural and educated – across generations and cultures. They write within diverse discourses of Iraqi women as having 'resilience' in Iraqi society and 'mother-like' strength (Kashou 2013). As story-tellers, Iraqi women are often configured as keepers of community remembrance, particularly in Shi'a Muslim communities (Deeb 2005; Pedersen 2014; Shabbar 2014). Other tropes in other contexts, such as Global North media outlets, present Iraqi women, conversely, as symbols of the oppressed and powerless in the face of war perpetuated by (militarised) male violence and 'long-standing' tensions between different identity groups. Alongside and despite such tropes, Iraqi women writers have always written their stories in a spirit of contra-punctual literary activism to any ideology that would situate violence, war, oppression and victimhood as categorically gendered. In this chapter, I focus on حبات النفتالين [*Mothballs*] by Alia Mamdouh and ما بعد الحب [*Beyond Love*] by Hadiya Hussein as examples of how some Iraqi women writers have sought to re/write the dynamics of confrontation in Iraq via Arabic and English para/translation. Mamdouh's novel حبات النفتالين [*Mothballs*] was the first novel by an Iraqi woman published twice in English translation after the outbreak of two wars in Iraq – once in the United Kingdom in 1995 and again in the United States in 2005. Hadiya Hussein's story ما بعد الحب [*Beyond Love*] (2003; 2012) focuses on bringing stories of the 1991 war to light across different times and places.

As noted by Nadje Al-Ali and Deborah Al-Najjar (2013), much Iraqi aesthetic production has been overshadowed by international political contexts of war, sanctions and violence in Iraq. In this chapter then, I explore how each work's translation confronts its own tensions of mediation through its *aesthetics* of mediation in relation to its deeply contextual politics of their meaning-making. To do so, I draw on feminist translation theories that seek to reconfigure translation beyond binary notions of replacement or 'loss' of meaning while exploring from an intersectional perspective how each novel expands and diversifies such theories' critical scope.

Beyond gendered binaries of violence: reading Iraqi women's literature of war and conflict

The politics of gendered confrontation, conflict and war has been a recurrent theme in many stories by Iraqi women writers. In the novel ممرات السكون [*The Corridors of Silence*], Iqbal Al-Qazwini (2006) writes of an Iraqi woman in Berlin vicariously 'living' the 1990–1991 Iraq war via the European TV screens and dreams, any categorical premise of herself – and other Iraqis in exile – 'not being there' and 'being there' thrown into question. In terms of the injustices enacted by time, Lutfiya Al-Dulaimi's novel حديقة حياة [*Hayat's Garden*] (2003) focuses on one woman, Hayat, living through the 1990 war in Iraq while waiting for her husband to return from the 1980–1988 war. Hayat's wait both ends and is tragically perpetuated when her husband is found to be alive – homeless in the streets of Baghdad, having lost his mind years ago when an ex-prisoner of war. She remembers *him*, but he does not remember her. In Bouthayna Al-Nasiri's short story "عودة الأسير" / "The Return of the Prisoner" (2000) the fallout of war in Iraq emerges in cross-generational and gendered frames. When a man – presumed martyred in the 1980–1988 war – returns to his wife and children, he does not receive a welcome as an ex-prisoner of war. When he discovers his 'death' as a martyr is much more palatable to his youngest son than his survival, he leaves the family for an unknown destination. This story shows that post-war conflict for Iraqi veterans is no longer the battlefield or the prison camp but the (gendered) society into which they return. Reading such stories brings up the question of how to read versions of Iraqi women's stories of conflict, when part of their politics and aesthetics is to project the richness and fabric of Arabic in different ways.

Feminist and activist translators have faced similar challenges when mediating asymmetrical post/colonial relations across languages in other contexts. Carolyn Shread (Flotow & Shread 2014) cites the difficulties of translating Fatima Gallaire's play *Les Co-épouses* (1990) into English from a French layered with Arabic and Amazigh, to 'confront' the limits of 'metropolitan' French as Algeria (and France's) sole language of mediation. Edward Said was also once told by a New York publisher that Arabic was "a controversial language" (1990, 31) when he proposed Arabic literary works for English translation. Clearly the limits Said was facing were not only due to Arabic as a language itself. So with the act of translation often an expression of movement across languages, Carolyn Shread (2007) has described translated works as "ethical encounters through exchanges in which difference is maintained within an intimate space" (2007, 213). Confines of censorship would be one example of such encounters. Stories moving across different receptions of a particular conflict, such as a war in Iraq, could be another.

To engage with such tensions of translation, Flotow and Shread have proposed a 'metramorphic' paradigm of feminist translation as one way of "refiguring the intimacy of translation beyond the metaphysics of loss" (2014, 592), an approach inspired by Bracha Ettinger (1992), an artist and academic working in the field of psychoanalysis. Amongst other things, earlier paradigms of psychoanalysis suggest that it is a person's sense of separateness and knowledge of their relation to others, the 'I' in relation to 'you' or 'others,' which gives rise to subjectivity formation. Sigmund Freud considered that it is memory, or the residue of memories, which helps us distinguish between feelings and thoughts and thus how "all knowledge has its origin in external perception" (1923, 23). Jacques Lacan also argued that human subjectivity, which includes a person's awareness and perception of her/his own 'self,' her/his sexuality and identity in relation to 'others,' emerges and is trapped through language (Lacan 1975). As an alternative to binary perceptions of 'One' and 'Other' manifesting through memory or language, Bracha Ettinger has proposed that we configure subjectivity-formation in less categorical terms by considering paradigms of the child/child-bearer relation. For Ettinger, this approach is not about reifying the biological capacity of women. She draws on this relation as one way of thinking through how human subjectivity can be understood as a co-emergent rather than confrontational process. A child-bearer carries a child-in-formation, while being-in-formation herself, often in uncomfortable spaces of mutual inter-dependency (Ettinger 1992, 200). Subjectivity-formation can then, in her view, be configured within frames of 'border-linking' of more than 'one' I and less than 'one' I ('I/non-I's): as subjectivities by which "elements or the subjects meet . . . recognise one another without knowing one another" (1992, 199). This notion of 'different-and-linked' suggests that different subjectivities could be in constant evolvement alongside each other without all elements of process being known by all involved. Such a paradigm helps us understand why creativity alongside 'limits' are vital sites of interrogation in relation to translation and why Flotow and Shread re/configure a metramorphic paradigm of feminist translation as an exploration of "the dynamics of intimacy . . . and hence the nature of translation itself" (Flotow & Shread 2014, 592).

How does metramorphic analytical framework of feminist translation work as a praxis? Shread (2007, 227) explains that it can involve engaging with a text in translation in ways "which do not efface its origins" (Shread 2007, 224). When translating Fatima Gallaire's play *Les Co-épouses* (1990), Shread sought to 'layer' the English version of the play in different ways (2007, 228). Ettinger's configuration of the Kabbalist notion of *tzimtzoum* – how God can make space for the world by contracting while remaining in the world – has been considered a helpful strategic critical frame to configure and interrogate dynamics of translation as processes of movement and co-emergent agencies operating in shared, confined, dangerous, sensitive and intimate spaces in different ways (Flotow & Shread 2014, 595). Questions on the ethical implications of reading stories of irresolvable conflict as generative however arise here. How can the ideologies of conflict interweaving Iraqi women's novels be read as co-emergent when the politics of US/Iraq conflicts, such as the 1991 Gulf War and the 2003 war? How could notions of border-linking be relevant analytical frames of reference for languages, such as Arabic and English, that are not mutually intelligible in script and grammar? With such critical questions in mind, I explore how the gendered politics of confrontation in حبات النفتالين [*Mothballs*] by Alia Mamdouh and ما بعد الحب [*Beyond Love*] by Hadiya Hussein move across languages. I first consider the extent that un/translatability of Mamdouh's novel حبات النفتالين [*Mothballs*] can be read as co-emergent, and generative when many crucial gendered aspects of her novel 'cannot' be read in English translation. I then explore how the politics of the 1991 uprisings in Iraq Hadiya Hussein's ما بعد الحب [*Beyond Love*] in Arabic are 'readable' as 'border-linked' to its version in English para/translation.

حبات النفتالين [*Mothballs*] by Alia Mamdouh (1986/2000): critical contexts

Born in 1944, Alia Mamdouh grew up in Baghdad and worked as a journalist before leaving Iraq in 1982. She has written seven novels and two collections of short stories. Her novel المحبوبات [*The Loved Ones*] (2004) was awarded the 2004 Naguib Mahfouz Medal for Literature, a prestigious award for the best contemporary novel published in Arabic yet to be translated. حبات النفتالين [*Mothballs*] (1986/2000) was one of her first novels, and it is still loved by many readers today. Najam Kadhim says that "تبقى برأينا واحدة من أجمل وأنضج الروايات النسوية والعراقية عموما لحد الآن " [it remains, in our opinion, one of the most beautiful and mature Iraqi and Iraqi *nisūwī* novels until this day] (cf. Al-Saffar 2018). Part of its appeal is how its main protagonist, Huda, evokes the sights, sounds and smells of Iraq of the 1940s by her distinctly Iraqi register. She tells of a life of poverty, childhood games, her defiance of her father, her fear for her mother dying and her love for her family. The politics of place – the family house, its courtyard, the narrow streets, the local mosque and neighbourhood shops – plays an important role in Huda's story of communities which now only exist as past local histories. As the title of the حبات النفتالين [*Mothballs*] suggests, this story is a "capsule of memory" (Faqir 1995, v), co-creating words which reconstruct

1940s Iraq. As Alia Mamdouh explains, many of her own memories are subject to change. In 1998, Mamdouh described her writing of Iraqi memory as a confrontation of a "multi-headed monster" (1998, 69): a monster which she termed as 'Arab fear,' its tentacles permeating everywhere. In her 2004 article, "Baghdad: These Cities Are Dying in our Arms," translated by Marilyn Booth, Mamdouh resituates her writing as an embodied political act of reconstructing Iraq in a post-2003 spirit of anger, not fear: "With a close-up shot, I am rebuilding my country before you. It is a construction of anger, vast beauty and identity which I submit to the act of writing. That will give it a finality" (2004, 48). Notable here is that Mamdouh does not specify to which language her re/construction of Iraq is submitted. What she does seem to specify, nonetheless is that the reader is as much part of this construction as she is.

حبات النفتالين [*Mothballs*] was first translated into English by Peter Theroux as *Mothballs* in 1995 as part of the Garnet Arab Women Writers' Series, edited by literary critic Fadia Faqir. After 1995, it was translated into many other languages. In her introduction, Faqir situates Mamdouh's novel in English as part of a transnational literary project which aimed to challenge "home-grown" and Western Orientalist ideologies projected onto Arab women/ writers before and after the 1990 war in Iraq (Faqir 1995, viii). On first page of *Mothballs* (1995), its outer cover, a reader is faced with an image of a public bath, a reference to a chapter where Huda is taken to the public baths by her aunt. Iraqi critic Mohammed Aref (2014) praises this very chapter in Mamdouh's story for evoking "حمّام النسوان يضاهي بسحره لوحة الفنان الفرنسي انغرز" المشهورة "الحمّام التركي" [images of public bath for women comparable in magicality to the painting "Turkish Bath" by celebrated French artist Ingres]. As Ingres' painting has been equally critiqued and pastiched for its Orientalising, patriarchal and voyeuristic overtones (Kleinfelder 2000), Aref's comment frames my question about the 1995 book and its potential epitexts of Orientalist exoticism towards women-only spaces in the Middle East. From a metramorphic perspective of translation, one way of considering this question is to look at *what* is actually presented as part of this work's meaning-making and what will co-emerge through reader engagement: co-emergence of meaning includes acknowledging elements of the unknown (Ettinger 1992, 199). The cover image appears from a child's-eye view. The bucket foregrounds the juncture where Huda, as a 9-year-old girl, starts playing around in the baths and causing chaos amongst women and children alike. The brush stroke images are of women talking or bathing, unconcerned with the outside world. Connected but separate to this image is the blurb cover flap which explains that the whole story is told "through the eyes of a nine-year-old Huda." As it lists the places framing the rhythm of Huda's life in Baghdad – public steam baths, roadside vendors and the streets where she plays and where political demonstrations take place –this blurb explains that the image is representing a chapter of the book, if 'a reader' takes the care and trouble to read it. In effect, this book can be read as 'facing' Orientalist representations of Arab women's spaces by c/overtly positioning readers as tasked with interrogating what they think this book is about from the outset. If a reader wishes to configure the cover page as resonating with Ingres' painting, it is up to them.

In 2005, the first English version was republished in the United States by New York Feminist Press as *Naphtalene: A Novel of Baghdad*. Its repackaging is striking. It has a new title, *Naphtalene: A Novel of Baghdad*, a choice which evokes flammable elements of 'mothballs' in a poetic way while carrying associations of Baghdad, a city synonymous in the US media with war and burning in post-2003 Iraq (cooke 2007). The first page, the outer cover, foregrounds a little girl wearing an *abaya* superimposed over a traditional Arab-world market scene. The front blurb cover flap then introduces Huda as the main protagonist and how she is "looking to establish her female identity amidst an oppressive patriarchal society and an impending revolution." It also explains how the novel is able to "change the way readers perceive gender politics in the Middle East." The back-blurb flap introduces Alia Mamdouh, Writer-in-Residence in Paris and winner of the Naguib Mahfouz Medal for Literature. The additional chapters of the 2005 version are equally striking. In her Foreword, philosopher Hélène Cixous describes the fieriness of Huda and who she is as a story-teller: "In appearance, a girl. In action, *a boy*. In poetic truth, *a fiery daughter*" (2005, vi) – her fire lighting up everything around her (2005, vi). The Afterword by literary critic Farida Abu-Haidar explains why Mamdouh's novel is such an iconic example of Iraqi and Arab women's literature and makes reference to the complexity of Mamdouh's use of Arabic: "her Arabic language often leaves the reader in confusion, and this of course is part of the appeal for many who love the ambiguities" (2005, 212). Abu-Haidar's explanation here and the sheer length of her Afterword suggest that 'the reader' is very much part of this story's English para/translation. It also suggests that a reader needs considerable guidance to figure out her/his part in this complex story.

Mothballs and *Naphtalene*: confronting gendered un/translatabilities and conflict in translation

Published in 1986, Mamdouh's novel حبات النفتالين [*Mothballs*] (1986/2000) literally defied inclusion, association with and co-option into any literature produced in Iraq or elsewhere at that time. It is a kaleidoscopic story of poetry, song, fable, magic realism, memory and gendered political satire which evokes different memories and associations of Iraqi culture at every turn. Eminent critic and writer Latifa Al-Zayyat (1995, 2) describes the magic of Mamdouh's story in poetic, co-emergent terms:

"يأتي الوضع باللغة العربية الفصحى وان امتزجت بالعامية العراقية. . . . بمدى ما يستثير هذا الوصف أجواء بغداد ولون ورائحة ونبض بغداد بمدى ما يبلغ استخدام الكاتبة للعامية العراقية مرتبة الشعر."

[It happens by Modern Standard Arabic and how it is intermixed with colloquial Iraqi . . . by the extent to which this type of description evokes the atmosphere of Baghdad, its colours, its smells and its pulse, by how the writer's use of colloquial Iraqi elevates and brings it to the level of poetry.]

(Al-Zayyat 1995, 2)

An example of the evocative power of colloquial poetic language referred to by Al-Zayyat can be seen here in an episode where Huda and her brother Adil are with their grandmother. Their grandmother tries to teach the children wisdom gathered from her long years of life through her sayings:

"اذا عملتو الخير لأحد لا تتحدّثوا به. ماذا يحدث في الحوش قولوا لا ندري. وإذا أودع أحد سره عندكم لا تفضحوه أبداً. السر مثل الكنز ولازم نخبئه بالبئر."

[If you do someone a good turn, don't tell anyone about it. Whatever happens here in the *ḥawsh*, say you know nothing. If someone confides in you with a secret, never let it out. A secret's like a treasure and we've got to keep it hidden away in a well.]

(1986/2000, 46)

Here we read a wise woman seeking to share her philosophy of life with her grandchildren in language that she thinks that they will understand. Her cadenza, sound and the choice of her words in Iraqi Arabic evoke a bygone generation of Iraqis whose diverse registers of Arabic undercut definitive divides between colloquial and formal Arabic expression in effortless ways. In the first sentence, the verb "عملتو" [*'amaltū*], a colloquial format of the second person plural, connects to the word "الخير" [*al-khīr*], which features in both colloquial and formal Arabic. It swiftly moves to the plural verb imperative "لا تتحدّثوا به" [*lā tataḥaddathū bihi*]: formal Arabic by the verb's vowel marking, yet destabilised by the less formal conjunction به [*bihi*]. Mamdouh seems to be 'attesting' here to what she thinks an (Arabic) reader can hear of this bygone generation's speech. She presents each register as not competing or overwriting the other, 'one' radically beautifying 'the other' without privileging, denying or conforming to 'the other.' Each register contracts to 'give space' to the other and creates *new* space where the critical impact of both registers can be expanded while blurring and subverting any 'official' borders between them.

As my back-translation attests, any attempt to render these words in English does not evoke the audible Arabic materiality in the same way. In both English versions of the novel, the meaning-making of the grandmother's voice is rendered semantically:

> "If you do a good deed for someone, don't talk about it. No matter what happens here at home, tell people 'We don't know.' If someone tells you his secret, don't ever repeat it. A secret is like a treasure and has to be hidden in a well."
>
> (1995, 33; 2005, 40)

Before we read the English versions as 'lacking' in relation to the Arabic, let us consider this excerpt from a metramorphic perspective of analysis. As Rosi Huhn (1993) explains, a metramorphic analytical approach configures notions of 'difference' and lack somewhat differently:

> in contrast to metamorphosis, each of the new forms and shapes of the metramorphosis does not send the nature of each of the preceding ones into oblivion or even eliminate it, but lets it shine through the transparency, disarranges and leads to an existence of multitude rather than unity.
>
> (1993, 8)

What of the creative audibility of the Arabic 'shines' through the English versions? The grandmother's rhythm of speech is still reiterated three times. Mamdouh's nuanced layers of inter-mixing Arabic are 'smoothed' into 'one' layer. Certainly, the audibility of the Arabic has much quieter echo in English para/translation. The intimacy of the connection between these two different generations of Iraqis in 1940s Iraq still remains. Huda still presents as 'creating space' for the advice of an older Iraqi woman to be co-emergently 'heard' by different readers across time. The point is to read the import of what is still there of the grandmother's wisdom rather than what is not.

One of the most challenging aspects of reading this novel, in whichever language read, is its themes of violence taking place before Huda's eye/I. Huda is regularly beaten by her father for her rebellious behaviour. Her brother Adil, clearly petrified by his father, regularly wets himself with fear. Huda's father evicts Huda's mother from the family house because he has married another woman, now pregnant, in the city of Karbala, where he works as a prison warden. As Huda – and he – find out, this violence does not go uncontested.

Her grandmother, as the family matriarch, confronts and resists his actions of patriarchy by never allowing the 'new family' into her family home – not out of revenge but to honour Huda and Adil's mother, a daughter-in-law she loves. Throughout the novel, patriarchy is 'outed' as a blunt and destructive tool of violence in all of its different permutations. Huda's paternal aunt, Aunt Farida, viciously attacks her husband physically and sexually for deserting her on the day of her marriage. She is not criticised for her attack on him but does not attain her goal of attaining a husband, a role she was brought up to expect, either. Huda notices how local Iraqi men in the neighbourhood expressing opposition to British colonial influence in Iraq 'disappear' and return with marks of torture. While this state-supported violence incapacitates many men on an individual level, it cannot quash the groundswell of movement against the British Empire influence in Iraq. But this is no 'horror' story about the patriarchal and colonial politics of 1940s Iraq. Alongside its many scenes of violence, Huda's eye/I witnesses other aspects of Iraqi life: humour, play and care between families, friendships between girls and boys. Women who love women make use of the many women-only spaces in the neighbourhood to have sex or simply enjoy cooking and conversation together. Huda's story evokes everyday life in this neighbourhood as a plethora of different cultures – cross-constituency confrontations *and* solidarities, often intersecting, challenging and transforming the effects on one other. This is perhaps what the 2005 blurb meant by the novel changing "the way readers perceive gender politics in the Middle East." But this does not explain why (US) English readers needed are provided with so many introductions and so much 'outer re-packaging' in order for their perceptions of gender politics in the Middle East to change.

As this point, potentially helpful here is revisiting what Abu-Haidar may have meant when she described the novel as a masterpiece of complexity (Abu-Haidar 2005, 212). From the novel's beginning, what the eye/I of Huda 'sees', co-emerges via Mamdouh using focalising techniques specific to Arabic. She co-creates diverse self-reflexive, gendered perspectives via grammatical structures which incorporate 'a reader' into its meaning-making. The story's first lines are emblematic of Mamdouh's technique:

"السحب فوق رأسك, والامتحان دائماً بانتظارك, انظري الى أبيك فقط تراءى لك أنه يقود شاحنة كبيرة, تجلس في الخلف أمك محتكرة الصمت والمرض, وباقي القطيع كان يلعب داخل المعتقل, يدمدم قليلاً ويسكت. جدتك . . . . "

[The clouds are over your head, and the test is always awaiting you. Look at your father. He looked to you like he is pretending/making himself look like he is driving a big truck. Your mother is sitting in the back monopolising the silence and illness. The rest of the herd play in the detention camp, growling a little then falling silent. Your grandmother. . . .]

(1986/2000, 7)

At this point, it is unclear what exactly 'the detention camp' refers to or where Huda and her family actually are. The reference to her father "تراءى لي" أنّه يقود شاحنة كبيرة [He makes himself look to me like he is driving a big truck] – is layered with different para/texts of ambiguity as the structure of Arabic verb "تراءى" [*tarā'ā*] (Wehr 1994, 368) communicates the action of seeing as reflexively and mutually carried out *between* people. This verb "تراءى" [*tarā'ā*] also implies that something is *feigned* – Huda's father being aware of his own (patriarchal) ridiculousness and fakery but continuing regardless. The word "انظري" [*anẓarī*] (1986/2000, 7) is the single feminine imperative of "look at, see" (Wehr 1994, 1144), which makes it clear a 'feminine' presence is the main point of address. The as-yet-unknown speaker is telling herself to look at herself *looking* at her father's (self) reflexive action – different to and apart from him while near and connected to him. While not immediately apparent at this point in the novel, the word تراءى [*tarā'ā*] is a crucial word. In Arabic, it is a relatively rare term, which makes it memorable, and it only appears twice: once at this juncture and then on the book's final page. In this way, the distinctiveness of تراءى [*tarā'ā*] in Arabic evokes and joins the story's end to where is started at its beginning – a girl looking at a patriarch 'feigning to drive,' with others in her family looking on:

"نركب الشاحنة التي **تراءى لي** أن أبي يقودها. عمتي في الدار الجديدة. جدتي تجلس بجوار السائق, ونحن نتمايل في الخلف."

[We are riding in the back of the truck which my father **looked to me as if he were pretending/feigning** to drive. My aunt is in the new house. My grandmother sits next to the driver and we sway in the back.]
(1986/2000, 212, emphasis added)

With the benefit of reading the story, a reader – helped by the second occurrence of the word تراءى [*tarā'ā*] – discovers that the truck 'mediated' by Huda at the very start of the story was in fact an actual means of vehicle transport all along. At the end of the novel, we understand that the family are leaving their neighbourhood due to the 1958 Iraq government slum clearances. This is not clear at the beginning of the story, as Huda does not seem to 'stay' in the truck for long. Her train of thought – like a film whose opening 'shots' are accompanied by no credits – very quickly moves on to 'see' other members of her family, followed by rooms in her house, family visitors and the streets on which she plays – catapulting the reader straight into the world in which she lives with little introduction. It is only after a number of pages that Huda, as a young Iraqi girl living in the Adhamiyya area of Baghdad, emerges as the mediator of this story. When تراءى [*tarā'ā*] features once more at the story's end, the clear resonance with the تراءى [*tarā'ā*] featuring in the first paragraph of the story helps a reader 'make the link' and so grasp the whole crux of the novel: the whole story is a stream of memory told by Huda, sitting in the back of a truck remembering her neighbourhood as she is leaving it. The word تراءى [*tarā'ā*] thus functions as a 'mothball of memory' to help the reader realise that s/he has been 'reading' a time-loop the whole time: Huda's eye/I re-seeing' her past life in 1940s and 1950s Iraq to relocate her memory of it into another eye/I – the reader's eye/I so as to widen its construction. تراءى [*tarā'ā*] thus expands where 1940s Baghdad can be read as being preserved in memory.

As the multivalent connotations of the verb تراءى [*tarā'ā*] are specific to Arabic, the two English versions 'preserve' this 'mothball' of Iraqi memory transference slightly differently:

> The clouds are over your head, and **the test is always waiting for you**. Just **look at** your father. It seems to you that he is driving a truck. Your mother is in the back monopolising the silence and illness.
>
> (1995, 1, emphasis added)

> The clouds are over your head, and **the trials of life are always ahead of** you. Just look at your father. It seems **to you** that he is driving a truck. Your mother is in the back monopolising the silence and illness.
>
> (2005, 1, emphasis added)

The 1995 version follows Huda's train of thought in Arabic: "the test is always waiting for you" (1995, 1). This sentence in the 2005 version is "the trials of life are always ahead of you" (2005, 1), a more explanatory approach. While subtle shifts in this sentence reveal (unnamed) para/translator editorship moving across the 1995 and 2005 English para/translation, the words "تراءى لك" [*tarā'ā laki*] are translated in both versions in the same way: "it seems to you." In the final paragraph, where the verb appears again تراءى لي [*tarā'ā lī* emphasis added], "it seemed to me" is used in both versions:

> As we rode in the truck it seemed to me my father was driving. My aunt was in the new house. My grandmother sat beside the driver and we swayed in the back.
>
> (1995, 162)

> As we rode in the truck it seemed to me my father was driving. My aunt was in the new house. My grandmother sat beside the driver and we swayed in the back.
>
> (2005, 190)

While a flawlessly idiomatic mediation of the Arabic, the reflexive capacity of تراءى [tarā'ā] to function as a potential 'mothball' of multi-focalised story-telling burns less brightly here, the politics of the novel's meaning-making dimmed or potentially 'lost' in English translation. The para/translators of the 2005 version seem to have noticed this risk and have tried to mitigate this risk via an Editor's Note:

> *Naphtalene* is narrated by Huda, the central character. Readers experience the various incidents and see the different characters through Huda's eyes. Yet *Naphtalene* is not an entirely "I" novel. . . . Mamdouh has chosen to use also the second person singular where Huda seems to be talking to herself. . . . There also occasional lapses into the third person. In this way readers have the impression they are witnessing events from different perspectives.
>
> (2005, viii)

Many publications are accompanied by editor notes, usually on a separate page before 'a story' begins. A reader then has the choice to read it or not. Here any chance of a reader 'missing' this Editor's Note is minimal – it directly *faces* the first page of the story (2005, ix) and literally 'spells out' that the novel is an 'experience' rather than a reading. A brief glossary of cultural terms (from the 1995 English version) orientates an English-language reader further. My question is: why does a reader need such 'step-by-step' guidance to approach this novel? The constant switch across 'I,' 'you' and other pronouns at the beginning of the novel is just as disorienting to read in English as they are in Arabic. After the first reference to the truck, for example, Huda's train of thought focuses on her grandmother, to then switch to Cousin Munir with all his distasteful habits, her Aunt Farida and then another aunt who shouts at "Huda" to "wipe the platter" (1995, 3; 2005, 4). Huda's younger brother, Adil, 'introduces' Huda to the reader, but via Huda's train of thought about him: "He loved you as if you were the last sister in the world" (1995, 4–5; 2005, 5). The link between "Huda" shouted at and the "you," the sister of Adil, only becomes completely clear once Huda's mother is mentioned: "She gave them Adil and Huda – what more do they want from her?" (1995, 5; 2005, 5). Until this point, a reader remains dis/orientated in Huda's world for five pages. Perhaps the para/translators of the English version were working to guide the reader to 'enter' and 'stick with' the story out of concern that s/he may not initially 'get' what this story is about and give up reading it. If this is the case, the very presence of such interventions uncannily reiterate something of this novel's self-reflexive dynamics: involving an/other as a crucial agent in its this story's meaning-making to ensure that a part of Iraq's social history is remembered by people other than Huda, or Mamdouh herself.

Intricately inter-weaving حبات النفتالين [*Mothballs*] in its English para/ translations, and specifically the 2005 version, are some assumptions: that 'a reader' cannot 'be present' without guidance. The para/translators of *Mothballs* (1995) and *Naphtalene: A Novel of Baghdad* (2005) may have been confronting a fear that this story may not be productively 'readable' around powerful para/texts of conflict and war surrounding Iraq at both times of its publication. Despite the pessimistic frames of such prem- ises, both English para/translations did have generative effects. The 1995 English version edited by Fadia Faqir led to the novel being published in other languages of translation. The Arabic version was reprinted by a Beirut publisher with wider-reaching distribution in 2000. The 2005 ver- sion also inspired New York Feminist Press to publish more novels by Iraqi women writers, and provide Iraqi women's literature with literary platforms of wider global reach. Wars in Iraq may have inspired some readerships' interest in Mamdouh's story in English. From a metramorphic analytical perspective of feminist translation, its gaps and supplementary interventions enact something of Mamdouh's politics of writing to con- front different frames of authority. Its English para/translations can be read as working to rebuild Iraq as Mamdouh did in Arabic, but before 'different eyes/I's.'

ما بعد الحب [*Beyond Love*] (2003): critical contexts

Hadiya Hussein is a novelist and short story writer with a long history of openly confronting political authorities through her literary activism. She was fired in 1995 from her job as cultural section editor for the Iraqi news- paper الجمهورية [*Al-Jumhūrīyya*] after publishing "A City of Silence" by Iraqi surrealist writer Adil Kamal (Šešić, Leštarić, & Alexander 2014, 103). She was blacklisted by the Iraqi Ba'athist government (Hussein cf. Lynx-Qualey 2013). She is one of the nine writers featuring in the recent anthology *12 Impossibles: Rebellious Arab Writers* (2014). Hussein herself came from a family who loved literature and song, and her father was a well-known local poet. Hussein believes that her writing has always been read as confronta- tional because she has always channelled her own creativity in ways which seek to preserve the beauty of Iraq and its local legends: "I write against the wars that distorted the beautiful features of my country" (Hussein cf. Lynx-Qualey 2013). Her novel ما بعد الحب [*Beyond Love*] (2003) was the first novel where she felt able to switch between two places – Jordan and Iraq – to write about features of her country fundamentally altered by the politics of exile (Al-Shabib 2014). It is a story about a woman named Huda who has fled to Amman after voting 'no' to Saddam Hussein in the local elec- tions and begins with Huda receiving the effects of her dead friend Nadia, killed in a car crash in Amman. Nadia's effects are little: a diary and letters to her lost lover, Emir, that were never delivered to him. The novel revolves around Huda spending most of her time in Amman reading this diary and letters, while listening and reading of other Iraqis telling their stories of international sanctions, poverty, war, rebellion and exile. They (and she) try to move on from the past but seem to talk about it endlessly.

The English version, *Beyond Love*, was published in 2012 by Syracuse University Press, an academic publisher renowned for promoting literature from many parts of the world. The (US) English version has striking additional para/texts which focus on two historical junctures of the 1990–1991 Iraq war and the 1991 uprising in southern Iraq: a Foreword by Arab literature scholar miriam cooke; an Introduction by translator Ikram Masmoudi; incisive reviews by eminent Middle East scholars Muhsin Al-Musawi and Roger Allen. The first page, the front cover, depicts buildings of Old Damascus by Syrian artist Emad Jano. Jano describes that he wanted to create an image representing "peaceful religious tolerance sadly lacking in present times" (Jano & Abou Rached 2017) – still somehow preserved in historical architecture of many cities despite the vicissitudes of different political eras. There are two episodes about the fallout of violence and intolerance that take centre stage in this novel: first, the fallout of the 1991 uprisings in southern Iraq, and second, a story of the 'Highway of Death.' Interweaving the English version is a strong para/text of shared authorship underpinning how these two episodes are mediated. In her Introduction, cooke presents Huda and Nadia as "creators whose survival challenges the destructiveness of war" (2012, x) – Huda in Amman, Nadia through her diary. Translator Ikram Masmoudi (2012, xvi) refers to Hussein's exile in Amman as mirroring Huda's own. On the outer cover, Iraqi academic scholar Muhsin Al-Musawi credits Hussein *and* Masmoudi as the translator for "unequivocally expos/ing the atrocity of American-led war and Saddam's revenge on the Iraqis. It is another testimonial in an inventory that should keep knocking at the human conscience." Roger Allen praises the novel for offering, "like many works of fiction . . . an understanding . . . that no news broadcast or journalistic report can replicate." Joining the English and Arabic version is the para/text of Iraq as a site of collective tragedy which needs to be read about by many different readers.

While there are many stories told or read in this novel, all are held together by Huda's experience of listening and reading *relaying* what she hears, reads and remembers. This is why the act of reading is such a key aspect of this story's meaning-making. Through Huda's own act of reading, many authorships (or aut/their/ships), known and unknown to each other, are brought together. Huda hears countless "flight from hell" stories in Amman at the UN Refugee Bureau (2003, 36) which have strong

emotive effects. When Huda hears an Iraqi parent call out to her child
"اسم الله يمّه" ["*ism allah yamma*"] in a park in Amman (2003, 133), she
'feels' this call as "thorns" in her heart (ibid). The accent reminds her that
many Iraqis carry their children as well as their 'stories' into exile, and the
sense that she has of constantly absorbing the many stories of Iraqi exile
through unconscious osmosis in ways also highlights the sheer numbers of
Iraqi exiles living in Amman and the connectivity of Iraqi experience. To
cope with this experience, we read how Huda makes choices on what stories she hears. She decides to take Nadia's diary and letters (2003, 7) rather
than leave them. At the UN office in Amman, she speaks to some Iraqis
and ignores others. When she meets Um Khadija in Amman, a woman
alongside whom Huda had worked in an underwear factory in Baghdad,
she seeks her out again to hear what happened to the other women in the
factory. Her landlord Sameh, a blind lute-player of Palestinian or Jordanian
descent, clearly wishes to tell his own stories of exile, but Huda seems
oblivious to them. Dominating Huda's time in Amman is one particular
story: the story of her friend Nadia, who she remembers through her diary
in ways which seem to confuse all boundaries between past and present and
who is telling whose story. Huda and Nadia met in Amman only to be separated by Nadia's death in a car crash. As the story of Huda reading shows
authorships co-emerging across the Arabic and English version in nuanced
ways, I focus my analysis on Huda's reading of Nadia's diary as aut/hered
by 'more-than' and 'less-than' one person. I also explore how Hussein herself enacts a politics of 'contraction' to allow other stories of Iraq to have
space to 'pass through' and so be read across different times and locations.

Para/translating confrontations of war: au/their/ing
Iraqi solidarity across languages

The first confrontation of aut/her/ship facing Huda is how and why she – as
'a reader' – comes to read the events in Nadia's diary. Nadia never wrote
her diary to be read or 'judged' by another reader, so nothing of what she
writes is 'sugar-coated' or euphemised to allow for reader sensitivities. Huda
would never have read the diary unless Nadia had tragically been killed. As
it is a diary, Huda thus cannot alter or 'change' anything of what she reads,
which is why she often feels pained when she reads it and often wonders
why she picked it up at all. She remembers one reason was boredom in exile:

"ماذا أفعل بالوقت المطوط؟ الساعات طويلة . . . لا شيء عندي أفعله. أيامي في عمّان
ساكنة مثل بركة لا تفضي إلى أي مجرى . . . ها هي نادية بعد موتها تحركها، تمسك بي
لأتسمّر أمام ذكرياتها."

(2003, 36)

WHAT COULD I DO with the lengthening hours? Time had slowed down. I had nothing to do. My days in Amman were quiet, like still water. But Nadia stirred it after her death, nailing me down in front of her memories.

(2012, 28)

Free time to read comes at a price. Whenever Huda tries to stop reading the diary, Nadia appears to her in dreams reminding her of her duty to 'read' it. Holding Nadia's diary is thus shown as a debt – a debt which can only be paid by reading it, even though Huda's first intention of reading was out of simple curiosity, a political dynamic of its meaning-making which extends to the politics of reading the novel itself. The second element of aut/hership facing 'a reader' of this diary are the sociopolitical contexts of which Nadia writes. She is Shi'a Muslim Iraqi woman born in a village and documents events specific to her and her family as Shi'a Iraqis. The excerpts in which Nadia first tells her story suggest the wider 'paratext' of Shi'a tragedy in Iraq never could (and should) have been open to change. Nadia notes, for example, that as she and her twin brother, Nadir, were being born, the midwife exhorted her mother:

"لا تخافي . . . إستعيني بالزهراء, أم الحسنين . . . . لا تشنّجي"

(2003, 34)

Don't be worried. Seek the help of Al-Zahra, the mother of the Hassanayn. Don't clench.

(2012, 27)

The name of Al-Zahra, a revered iconic figure in Shi'a Muslim history, sets a highly politicised dynamic of survival for Nadia and her twin brother that is predicated on historicised discourses of gender: men's fate as martyrdom and women's roles to commemorate past injustices as part of, not separate from, those of the present (Shabbar 2014, 212). As the mother of Imam Hussein, killed in battle with his twin brother Hassan at the Battle of Karbala, Al-Zahra is invoked in Shi'a commemorations for two reasons: one, as a mother who has lost her sons to the armies of Yazid, the Sunna Muslim Caliph, and two, as a mother to a daughter, Zainab, a figurehead revered for keeping men safe from the armies of Yazid (ibid). As Nadia's mother had a history of miscarrying baby boys but bearing healthy girls (Hussein 2003, 34), the midwife's invocation of Al-Zahra is to prepare her for the death of yet another boy. To everyone's surprise, two babies are born: not two boys, but a girl followed by a boy, a gendered transformation of the Hassanayn (the twin boys Hussein and Hassan) narrative, boy and girl horizontally side by side in timeline, not vertically separated by generation. This event shows that death happening in the past cannot, does not and should not foreshadow death in the future, which is what Nadia's diary – read in Arabic or English by Huda or whoever is reading the novel – seems to be about.

The ambiguous location of Huda and the reader reading Nadia's diary as a document of actual lived experience makes the borders between Huda's memory and Nadia's writing porous in the Arabic version and the English version in different ways. Huda remembers how Nadia was reluctant to talk about her about how her twin brother, Nadir, ended up dying before her after all: he was executed by the Iraqi state authorities due to taking part in the 1991 uprisings against the Iraqi Ba'athist government in Basra. Nadia's reluctance is partly because the experiences were too painful to recount and also because she felt that her own story was not worth repeating: her story is the same as so many other Iraqi stories.

"المصيبة اننا نحكي الحكايات ذاتها مع علمنا ان كل عراقي عاشها واكتوى بنارها".

[The catastrophe (of it all) is that we tell exactly the same stories despite all of us knowing that every Iraqi person has lived them and been charred by their fire.]

(2003, 25)

The irony is that we all recount the same stories even though we know that every Iraqi has been burned by this fire.

(2012, 18)

Once again, Nadia situates her tragedy as part of wider dynamics of oppression and violence affecting everyone in Iraq, regardless of cultural, religious and political identity. Huda replies that Nadia needs to tell more of how her twin brother, Nadir, was executed by the Iraqi authorities for her own sake:

"ليس لنا الا هذه الحكايات, علينا أن نكررها لتبقى شاهداً على عصر المجازر . . . تكلمي يا نادية . . . أخبريني لماذا أعدموا نادر؟"

[We have nothing else but these stories, so we must repeat them so that they (the stories) are a witness to the era of butcheries. . . . Nadia, tell me . . . why did they put Nadir to death?]

(2003, 25)

Huda is urging Nadia to tell her so that she and Nadia can mourn him together, which they do after Nadia tells her story. It is the stories of death, rather than the people in them, who act as "شاهداً" [a witness] (2003, 25). Stories *about* them are the only remains left of them, carried by others still alive. Carried by Nadia and Huda in this scene, Nadir co-emerges with an uncanny 'double-presence' recalled by Nadia with Huda while Nadia was still alive; *and* once again remembered *by* Huda recalling Nadia telling her *about* Nadir after she had died. This juncture in the novel configures space for the story of Nadir (and those suffering similar fates) to have a presence via the dynamics of writing and reading.

In the English version, however, Huda's call on Nadia to tell the story of Nadir is imbibed with a slightly more *public* politics of empathy and solidarity-building which subtly expands and diversifies the political para/ text of the sentence for (US) English-language readerships:

> "These stories are all that we have. We ought to bear witness to the age of butcheries. You have to speak, Nadia. Tell me why they put Nadir to death."
>
> (2012, 18–19)

Here the action of witnessing shifts to 'we' – whoever 'we' are – instead of the stories themselves. a subtle shift which configures the act of *people* listening alongside the act of speaking as generative *and* co-emergent activist action. Through Huda's act of reading more about Nadir in Nadia's diary, she/ we find that Huda's belief in the power of people speaking and listening as a moment of witnessing has some limitations. When Huda reads Nadia's diary entry about Nadir's death, Huda discovers that Nadia did not 'speak' the 'full story' of what happened to him in 'the age of butcheries.' She wrote it in a place where she believed no one else would ever see/read it – her diary. She writes details of horrors that she saw which she may have not have shared when speaking: such as how hundreds of other Shi'a families, like her and her mother, had to collect coffins of their executed loved ones from the Iraqi prison authorities in silence to take them to Najaf cemetery, a large burial ground for Shi'a Muslims in Iraq (2003, 61; 2012, 56). She also describes their journey to Najaf cemetery and what happened there after she and her mother discovered to their horror that the coffin did not contain Nadir's body but the corpse of a much older man, his body mutilated and eyes gouged out (2003, 63; 2012, 57). This discovery leads to other mourners opening their loved ones' coffins, too, all finding the mutilated bodies of their loved ones – at times body parts – in the wrong coffin. One man even finds that one coffin had three, not two, legs (2003, 66; 2012, 60). In her diary, Nadia asks herself unspeakable questions which nonetheless bear (silent) witness to how Nadir – and other Iraqis – may have been tortured:

"كيف خرجت روحك يا توأم روحي؟ كم لحظة دامت نفسك الأخير؟ . . . اخسى ان يكونوا قد قطعوا جسدك وانت حي . . . وين سؤال وسؤال تصرخ أعماقي: أين اختفت جثة نادر؟"

(2003, 66)

In English, the same moment of pain is recorded, with some slight altera-
tions of font format:

> *How did your soul depart, my twin? How many moments did your last
> breath linger? I'm afraid they dismembered your body while you were
> still alive.*
>
> Between one question and another, my soul was screaming: "Where
> is Nadir's body?"

(2012, 60)

This is a scene where personal psychic pain of one woman is shown as
relatable to many others' experience. Her expression of the pain is a pain
undoubtedly felt by all of the mourners at that cemetery and by anyone
else – Iraqi or otherwise – finding out that a loved one has been tortured. As
the personal depth of her own experience is particular to her, this is why, as
stated by Huda earlier, each story needs to be told.

A metramorphic analytical approach of feminist para/translation could help
us focus on the importance of para/textual interventions in showing how the
political dynamics of this horror described in this diary in Arabic and English co-
emerge alongside each other but differently. In the Arabic version, each thought
seems to run into the other, line after line. In English, there is a break between
the first trains of questioning and the last question signalled by a new indent
and change of font, all of which configure Nadia's final question as potentially
'apart' or separate from while remaining connected to the others preceding it.
The lines in italics are Nadia's chain of thought. But the unexplained change of
font and break in the final line creates an interesting ambivalence of agency and
so raises questions on whether a potential change of agency has taken place: is
this final question still Nadia's train of thought? Or is it Huda's own question in
response to Nadia's first questions? Reflecting on this intervention highlights,
once again, the importance of reading what appear to be innocuous para/textual
mediations as an echo of what we may miss in this novel's complex web of
co-emergent meaning-making. Here the para/textual shifts enact Nadia's lines
of questioning as open to be fractured and shared by others. This suggests that
the acts of writing, reading and reflection are border-linked activities happening across more than one person. In other words, this expression of pain is split
and joined in agency. Subtly foregrounding what is implicit in the Arabic, the
English version shows how the collective and individual specificity of the torture meted out on her brother and others after the 1991 uprisings come to light
through many, not just one, mediating agencies.

Many of the entries about events at the Najaf cemetery in Nadia's diary
initially seem to represent a specific Iraqi Shi'a experience of death in
1991 Iraq as connected to wider Shi'a collective memory (Deeb 2005, 15;

Shabbar 2014). In this novel, this memory seems intertwined with formal Arabic and English as two languages of power enforcing and mediating the collective tragedy of many. Formal Arabic is the language by which the (Iraqi Ba'athist) government occupied and exercised authority over the Iraqi public sphere, which included the banning of many cultural religious practices relating to Shi'a and 'non-conforming' Iraqi identities in general (Ali 2008). The oppressive silence pervading the mourners collecting the coffins from the Iraqi authorities testifies to this, as do the Iraqi state officials as they read out the roll call of executed persons by adding the prefix "الخاين" or "the traitor" (2003, 58; 2012, 50) to each name. English is implicated in this collective memory of mass killings, too, although it is never articulated in Nadia's diary. Huda reads of how thousands of Shi'a Muslim Iraqis – children, mothers and elderly – in southern Iraq were massacred *en masse* by the Iraqi Republican Guard forces during the 1991 Basra uprising (2003, 81; 2012, 76) – actual historical events well documented by many international organisations at the time (Human Rights Watch 1992). In Nadia's diary and her letters to her lost lover, Huda – and the reader – Huda reads in much more localised detail how the Iraqi Republican Guard forces carried out the killings. She (and readers) also read how US military helicopters, their chains-of-command mediated via US English, were present, too: remaining hovering in the skies, silently watching massacres taking place (2003, 81; 2012, 76).

Considering the reading of this event as a representation of co-emergent agencies of 'less than one' yet 'more than one' subjectivity (Ettinger 1992, 199) can help us to understand the politics of Nadia's diary moving across languages. For Huda, the para/text of Nadia's diary is that it is not a story: it is Nadia documenting what she and other Iraqis experienced at a particular juncture of time and location in Iraq. Her focus is specific to the 1991 uprising in southern Iraq. Uprisings, after all, happened elsewhere in Iraq, with similarly retributive actions by Iraqi Republican forces (Al-Jabbar 1992, 9), but Nadia did not experience them directly. Perhaps she did not even know about them at the time. Through considering the genres of writing, that is, a diary and a novel, as 'borderlinking' different subjectivities of 'lived experience' and subjectivities represented through fiction, the Shi'a specificity of this particular event evokes or suggests potential connections to other these other events of persecution taking place in other regions of Iraq, many of which have not been recognised and discussed (Hussein cf. Lynx-Qualey 2013) but without Nadia 'knowing' them to write about them. In this way, the telling of this story of Basra is a call for witnesses of these other events to write their own stories. In English para/translation, the borderlinking of diary and novel as two different and overlapping genres of story-writing expands these stories' political range by situating the act of reading about such events as proxy for the act of 'witnessing' them across languages.

Nadia's diary is not the only instance or site which brings to light the politics of reading stories in this way. Another radical instance of reading as an act of solidarity co-emerges through *another* diary, handed to Huda in the form of a letter by an Iraqi man, Moosa, who she meets in Amman. After Huda's political asylum case is refused by the US Embassy, Moosa offers Huda the chance to travel to Australia with him on his family card, as his case has been accepted. Moosa gives her a letter written by him for two reasons: because he wants to explain something about his past life in Iraq and because he believes that future generations need to know his story (2004, 139; 2012, 130). The letter is about his time as a soldier during the 1990–1991 Gulf War and his experience on the 'Highway of Death,' a route from Kuwait to Basra on which many Iraqi soldiers returned after Iraq's defeat in 1991. In this letter, he details how the Iraq ground troops on the way to Basra were bombed out by US military air forces before the 1991 cease-fire came into effect – literally tearing thousands of men limb from limb. As with Nadia's diary, the politics of reading this letter in Arabic or English cannot be configured as a neutral act: the graphicness of the descriptions of how individual Iraqi soldiers die at the hands of the US military make this letter very difficult reading. Such descriptions, however, give presence to many unknown Iraqi soldiers, many of whose bodies may have simply evaporated from the force of weapons deployed by the US military air forces. It also gives discomforting context to why the US military forces and affiliated administrations in 2003 Iraq were met by many Iraqis with ambivalence, despite the Iraqi Ba'athist government's long history of atrocities (Cahill 2016; Chediac 2016). The biggest shock for Moosa is not how his compatriots died but rather the very precise timing of their killing:

"حين وصلت إلى تلك المجموعات عرفت أن الثامنة صباحاً – اي بعد دقيقتين – ستكون
... موعداً لوقف إطلاق النار!! ... كيف سمحوا لأنفسهم بقتلنا قبل خمس دقائق لا غير
خمس دقائق كانت تفصل فتية كثيرين عن الحياة؟"

[As I reached those (military) units, I learned that at 8.00am – that is, in two minutes' time – would be the ceasefire!! . . . How did they allow themselves to kill us five minutes ago . . . five minutes had separated so many young men from life?]

(2003, 148)

His question holds many ramifications: how indeed could the US pilots bring or 'allow' themselves to slaughter so many Iraqis at this time? He does not respond to his question, as the reason is obvious: the US pilots were following orders issued by their chain of command just as they, as

Iraqi (conscripted) soldiers, were ordered to occupy Kuwait by theirs. The English version reads slightly differently:

> How could they have been killing us only five minutes earlier?
> (2012, 139)

This re-focalising translation strategy reframes the question: how could 'they' *have been* doing it rather than *allowing themselves* to do it? The shift with/holds the soldier's rhetoricity in Arabic to carefully resituate what has to be read about. As expressed in the words of Musa Al-Musawi on the back cover: "the atrocity of American-led war and Saddam's revenge on Iraqis." This shift contracts the rhetoricity to co-create a space in which an Iraqi soldier – identifiable as a potential perpetrator of violence – interrogates other perpetrators of violence: how did 'they' – the US military forces – come to 'be' mass killing at this exact time? Questions on why perpetrators of violence are in certain places at very specific times at times of war goes much further than the 1991 war in Iraq itself. US war photographer Harold Evans recalls that it is only a picture of death and traces of destruction in war in the media which reminds us that "We have willed it by sending the soldier there to do that dirty work for us" (cf. Carr 2003). As many pictures relating to the US military in Iraq were deemed as unpublishable in the US media during and after the 1990–1991 Iraq war (DeGhett 2014), this letter co/emerges as one of the most detailed *discursive* pictures of the Highway of Death readable in English. Along with Huda – and Arabic-language readers – reading the letter in Arabic, this letter in English para/translation thus tasks its 'reader' with confronting her/his position on why the US were there killing Iraqi soldiers at that particular time.

The radically co-emergent politics of this letter appearing in this novel – and Nadia's diary – in both languages do not stop at this point. The letter initially presents as 'written by' Moosa (a fictional character) via Hadiya Hussein (a real-world person) which is received and 'read by' Huda (a fictional character) and an unknown (real-world) reader. In English para/translation, Ikram Masmoudi mediates Hussein's writing from Arabic into English. At the end of this 'letter' from Moosa, an unannounced para/textual 'twist' occurs: a 'real-world' reader is addressed directly – via an asterisk – but not Huda. In both the Arabic and English version, this asterisk signals a footnote: "from a diary of an Iraqi soldier, poet Ali Abd El-Amir, dated March 2nd, 1991" (2003, 148; 2012, 140). As Ali Abd El-Emir is a 'real-world' Iraqi poet, this footnote represents a significant juncture in the novel. It first overtly 'presences' a reader as part of the novel's meaning-making in both Arabic and English para/translation. And whether it is Abd El-Emir's 'real' memoir or an autobiographical fiction written by him, this asterisk secondly signals the presence of a rare genre of Iraqi writing: 1991 war writing, far from the stories glorifying war published about the 1980–1988 Iran-Iraq war during the era of *Qadisyat Saddam* (Khoury 2013; Lewental 2014). This asterisk also para/translates in both versions of the novel the radical power of authorships (and translatorships) which 'contract' in *tzimszum*-like ways (Flotow & Shread 2014, 595) in order to make room for others. Hussein 'contracts' her aut/her/ial space to make room for Abd El-Emir. In 'the house of fiction,' Huda contracts her own thoughts to 'read' the letter whose contents co-emerge on the page by her act of reading it. Her contraction does not occur without purpose – Moosa has offered take Huda to Australia in his UN travel card by using his leverage as an 'accepted' refugee to create space for a 'refused' refugee to 'pass through' the unsurmountable barriers created by interlocking dynamics of immigration. In response, Huda's act of reading 'lets pass through' a story which she – as a non-military Iraqi woman – could never write herself. Hussein and Al-Amir also contract their au/their/ships: to let pass the possibility that 1991 war stories, however few in number, unpublished or unwritten, could and do exist, by 'ciphering' this letter as co-emergent in authorship via in authorship the agencies of Huda (reading the letter) and Moosa (writing it) in 'the house of fiction.' Whether a 'true' account of events on 'the Highway of Death' occurring on the 2 March 1991 or a fictional representation, the political importance of this account existing in both versions of the novel is significant: for the first time, *something* of what happened to Iraqi soldiers on the Highway of Death in 1991 can be 'seen' as written *and* 'read' via this 'letter' being given space by several authors, or 'autheirs' passing through a novel published in Arabic and English para/translation.

Considering the novel as a work of au/their/ships 'letting pass' other au/their/ships helps us appreciate the generative political intent of cooke, Masmoudi, Al-Musawi and Allen as para/translators of this work in English. The ISBN page of the English version – a 'routine' list of publication information and copyright details – states the following: "This is a work of fiction. Names, characters, places, dialogues and incidents are either the product of the author's imaginations or are used fictitiously. Any resemblance to actual persons living or dead . . . events, or locales is entirely coincidental." Although 'routinely banal' in nature, an ISBN note presenting all events in the novel as fiction creates an uncannily ambivalent status for this novel: is it a memoir or a work of fiction? In contrast to the ISBN page, the para/translators of the English version ensure that events represented in Hussein's novel are read as a 'testimony' to actual events in Iraq which call to be witnessed in the world of the real, not of the imaginary. In the words of reviewer Muhsin Al-Musawi, one of the para/translators of the outer back cover: "It is another testimonial in an inventory that should keep knocking at the human conscience." Here he configures 'testimonial' as a list which has an almost physical and personal agency, and the human conscience as a door to political action. Roger Allen's review on the back cover which states that Hussein's story can tell us more than any news report also squarely frames the events in this story as not a product of Hussein's imagination, nor used by her fictitiously. From a metramorphic analytical perspective of feminist translation, the reading of this novel as a testimony as 'border-linked' – one which is 'passing' via – while not fully belonging to – the genre of fiction helps understand why the act of reading it, in light of these commentaries, is as a generative political act.

Reading the novel as testimonial cannot change the tragic outcomes of any of its stories in Arabic or in English para/translation. As a novel about the deaths of thousands in Iraq, no alternative outcomes or 'happy endings' are possible. The decision to read these stories in either or both languages, however, creates potential space for changing the outcomes for how people and their tragedies can be understood and remembered. No monument, for example, was ever erected to 1991 war survivors in Iraq – partly because many of its survivors, like Moosa, joined the 1991 uprising and were later killed. Moosa/El-Emir's letter can, thus can be read as an alternative 'monument' to the many deaths, injuries and survivals *known* and *unknown* on the Highway of Death. Nadia's diary similarly accounts for many people in prisons, the many relatives not allowed to bury and mourn their dead, and many others quietly living personal tragedies of war and exile by choosing not to speak. The novel is a testimony to many known and unknown Iraqi stories moving across borders in Arabic and English. Via its (self) reflexive modes of mediation, each version of Hussein's novel enacts how words and the act of reading can function as retrospective edifices to the absent/ed.

Ongoing questions

Re/reading Iraq women's stories in English (para)translation

Iraqi literature has survived and thrived despite many Iraqi writers having to negotiate highly charged, gendered, political discourses of censorship of solidarity (Alsagaff 2018). This helps explain why many Iraqi women writers have worked to 're-write' or 'recover' Iraq as an ever-emergent place of identity going far beyond any geographical borders 'decided' by hegemonic spheres of authority. The uncanny imaginaries threading through the works by Samira Al-Mana (1990; 2005) and Daizy Al-Amir (1988; 1994) mark opacity in para/translation as a crucial politics of engagement which merits more in-depth exploration and research. Inaam Kachachi's story of an Iraqi-American woman translator in post-2003 Iraq (2009a; 2011) raises questions on the 'audibility' of Iraqi literary and cultural production: whose voices do we think we are 'hearing' or 'reading' when experiencing a literary work in English para/translation? Betool Khedairi's polyphonic story of Iraq (1999; 2001) confronts us with the tensions of Iraqi writers representing the most marginalised social groups of Iraq across different languages, times and places via the medium of writing. If we reconfigure such tensions as points of interrogation, we can celebrate and recognise that the many instances of Iraqi women's story-making, such as Alia Mamdouh's حبات النفتالين [*Mothballs/Naphtalene*] (1986/2000), which 'defy' translation as instances of re/writing calling for ongoing critical engagement. Other Iraqi women's stories just call to be read in whatever language they are published- Hadiya Hussein's ما بعد الحب [*Beyond Love*] (2003; 2012) is just one example. As this book has shown, Iraqi women writers did not negotiate their politics of translation single-handedly. Other para/translators were part of their pathways of mediation. As Susanne De Lotbinière-Harwood (1991), citing Luce Irigaray, reminds us: "Écrire n'est jamais neutre. Traduire non plus" [Writing is never neutral. Neither is translation] (1991, 11). This book on reading Iraqi women's stories has shown, the act of para/translation is not either.

Iraqi women writers' commitment to mediate stories of Iraq and Iraqis in counter-hegemonic ways have called for varying pathways of para/translation, many of which 'co-create' (while being co-created) by constantly evolving receptions and environments. Haifa Zangana, for example, recalls how one woman thanked her for telling the story of a friend of hers in her novel *Women on a Journey* (2014, ix) – someone she did not know at all. For Zangana and many other Iraqi women writers, such encounters are of no surprise: "some books like their characters, just have a life of their own" (Zangana & Abou Rached 2018). For many with memories of Iraq, like this woman, the borders between lived reality and fiction very much have their own lives too. This is why many Iraqi women's stories often carry such a great emotional as well as political impact whether read in Arabic or English para/translation. For this reason, we find that any engagement with Iraqi creativity implicates the reader, including those reading, teaching and commenting on their works, often in various languages. The stories that Iraqi women tell call on their readers to consider the multiple times, places and texts in which Iraqis are working to preserve and rebuild understandings of Iraq and their identities in different ways.

The gendered politics and aesthetics of Iraqi women's stories moving across borders, times and locations have been the primary focus of this book, with translation one point of departure. While this book is the first to focus on Arabic-English translation of Iraqi women writers' story-making using analytical perspectives of feminist translation studies, its critical relevance extends to literary and cultural traditions in other contexts. A novel, play, film or a story, for example, may well be the only written record of localised histories, landscapes and customs erased from official landscapes. Iraq story-writing is one instance of how literature can preserve the memories of many, This book has proposed new ways by which stories in diverse formats and languages can be read as vital documentations of places, peoples and events and has shown how analytical frameworks of intersectional feminist translation studies can helps us read such stories beyond derivative, or binary frames of translational comparison. This book does not, however, set out any categorical theoretical agenda for how expressions of para/translatory activism can or should be read. As Ferial Ghazoul points out, much of Iraqi story-writing must be read as "an aesthetic expression of a complex and disturbing reality" (2004, 1). This reality includes the languages and ways in which Iraqi writers have published, and in many cases, continue to publish. This book has focused on pathways of English translation as just one way of appreciating the realities and imaginaries that Iraqi women writers are working to transform through their story-writing across many times, situations and places.

Bibliography

Primary sources (in order of analysis)

Chapter 2: Translating 'the uncanny': Samira Al-Mana and Daizy Al-Amir

Al-Amir, Daizy. 1988. على لائحة الانتظار [*On the Waiting List*]. Beirut: Dār al-Ādāb.

Al-Amir, Daizy. 1989/1994. "Author's Preface." In *The Waiting List: An Iraqi Woman's Tales of Alienation* by Daizy Al-Amir, tr. Barbara Parmenter, ix–xii. Austin: Texas University Press in Austin.

Al-Amir, Daizy. 1994. *The Waiting List: An Iraqi Woman's Tales of Alienation*, tr. Barbara Parmenter. Introduction by Mona Mikhail. Austin: Texas University Press in Austin.

Al-Mana, Samira. 1990. حبل السرّة [*The Umbilical Cord*]. London: Al-Ightirāb al-Adabī.

Al-Mana, Samira. 2005. *The Umbilical Cord*, tr. Samira Al-Mana and ed. Charles Lewis. Yorkshire/London: Central Publishing Services/Al-Ightirāb al-Adabī [Literature in Exile].

Chapter 3: Translating gendered dis/location in post-2003 Iraq: Inaam Kachachi

Kachachi, Inaam. 2009a. الحفيدة الاميركية [*The American Granddaughter*]. Beirut: Al-Jadīd.

Kachachi, Inaam. 2009b. *If I forget you Baghdad, The American Granddaughter*, tr. William Hutchins [Online]. http://intranslation.brooklynrail.org/arabic/if-i-forget-you-baghdad-by-inaam-kachachi [Accessed 10th April 2020]

Kachachi, Inaam. 2010. *Se je t'oublie Bagdad* [*If I Forget You Baghdad*], tr. Khalid Osman. Paris: Éditions Liana Levi.

Kachachi, Inaam. 2011. *The American Granddaughter*, tr. Nariman Youssef. Doha: Bloomsbury Qatar Foundation Publishing.

Chapter 4: Conversations about 'solidarity among the subaltern': Betool Khedairi

Khedairi, Betool. 1999. ‏كم بدت السماء قريبة!‏ [*How Close The Sky Seemed!*]. Beirut: Al-Mū'asissa al-'Arabiyya Li al-Dirasāt Wa al-Nashr.
Khedairi, Betool. 2001. *A Sky So Close*, tr. Muhayman Jamil. New York: Pantheon Books.

Chapter 5: Re/writing confrontations in translation: Alia Mamdouh and Hadiya Hussein

Hussein, Hadiya. 2003. ‏ما بعد الحب‏ [*Beyond Love*]. Beirut: Al-Mū'asissa al-'Arabīyya Li al-Dirasāt Wa al-Nashr.
Hussein, Hadiya. 2012. *Beyond Love*, tr. Ikram Masmoudi. Preface by miriam cooke. Introduction by Ikram Masmoudi. New York: Syracuse Press.
Mamdouh, Alia. 1986/2000. ‏حبات النفتالين‏ [*Mothballs*]. Beirut: Dār al-Ādāb.
Mamdouh, Alia. 1995. *Mothballs*, tr. Peter Theroux. Preface by Fadia Faqir. Arab Women Writers' Series, ed. Fadia Faqir. Reading: Garnet Publishing.
Mamdouh, Alia. 2005. *Naphtalene: A Novel of Baghdad*, tr. Peter Theroux. Foreword by Hélène Cixous. Afterword by Farida Abu-Haidar. New York: Feminist Press.

Secondary sources

Abd El-Amir, Ali. 1992. ‏يوميات جندي عراقي عائد من الهزيمة بعد تحرير الكويت‏ "[Memoirs of an Iraqi Soldier Returning from Defeat After the Liberation of Kuwait] (Dated 2nd March 1991)." In ‏ما بعد الحب‏ [*Beyond Love*] by Hadiya Hussein, 2004, 140–148. Beirut: Al-Mū'asissa al-'Arabiyya Li al-Dirasāt Wa al-Nashr.
Abdel Nasser, Tahia. 2018. "Dreaming of Solitude: Haifa Zangana and Alia Mamdouh." In *Literary Autobiography and Arab National Struggles*, 110–129. Edinburgh: Edinburgh University Press.
AbdelRahman, Fadwa K. 2012. "Writing the Self/Writing the Other in Thomas Keneally's *The Tyrant's Novel* and Inaam Kachachi's *The American Granddaughter*." *Postcolonial Text* 7, no. 3: 1–19.
Abdullah, Angham A. 2018. "The Inexpressible in Iraqi Women's Narratives of War." *The Feminist Journal of the Centre for Women's Studies* no. 1. https://cultivatefeminism.com/protests-the-inexpressible-in-iraqi-womens-narratives-of-war/ [Accessed 10th April 2020].
Abou Rached, Ruth. 2017. "Feminist Paratranslation as Literary Activism: Haifa Zangana in Post-2003 America." In *Feminist Translation Studies: Local and Transnational Perspectives*, eds. Olga Castro & Emek Ergun, 195–207. London/New York: Routledge.
Abou Rached, Ruth. 2018. "Many Women on Many Journeys: Haifa Zangana's *Women on A Journey: Between Baghdad and London*." War and Occupation in Iraq: Women's Voices, Gender Realities, eds. Brinda Mehta & Haifa Zangana. *International Journal of Contemporary Iraqi Studies* 12, no. 1: 51–72.

Abou Rached, Ruth. 2019. "Remembering the Literary Achievements of Daizy Al-Amir." *Journal of Contemporary Iraq and the Arab World* 13, no. 2–3: 271–287.

Abou Rached, Ruth. 2020. "Pathways of Solidarity in Transit: Iraqi Women's Story-making in English Translation." In *The Routledge Handbook of Translation, Feminism and Gender*, eds. Luise von Flotow & Hala Kamal. New York/London: Routledge, 48–63.

Abu Elhija, Dua'a. 2014. "A New Writing System? Developing Orthographies for Writing Arabic Dialects in Electronic Media." *Writing Systems Research* 6, no. 2: 190–214.

Abu-Haidar, Ferida. 2005. "Afterword." In *Naphtalene: A Novel of Baghdad*, by Alia Mamdouh, tr. Peter Theroux, 191–213. New York: Feminist Press.

Ahmad, Hadil A. 2017. الرواية النسوية خارج فضاء الوطن: روايات عالية ممدوح انموجا [*The Nisūwī Novel: Outside Spaces of the Homeland: The Novels of Alia Mamdouh as a Case Study*]. Amman: Dār al-Ghaida'a. [E-Publisher: Al-Manhal]

Al-Ali, Nadje S. 2007. *Iraqi Women: Untold Stories from 1948 to the Present*. London: Zed Books.

Al-Ali, Nadje S., & Deborah Al-Najjar. 2013. *We Are Iraqis: Aesthetics and Politics in a Time of War*, eds. Nadje S Al-Ali & Deborah Al-Najjar, xxv–xl. New York: Syracuse University Press.

Al-Ali, Nadje S., & Nicola Pratt. 2009. *What Kind of Liberation? Women and the Occupation of Iraq*. Berkeley: University of California Press.

Al-Amir, Daizy. 1968. ملتقى النهرين"[Where the Two Rivers Meet]." *Al-Ādab*, March, 34–35, ed. Suheil Idris. https://Al Ādāb.com/sites/default/files/aladab_1968_v16_03_0034_0035.pdf [Accessed 10th April 2020].

Al-Amir, Daizy 1992. 'إبداعية نسوية شهادات' ('Creative Women's Testimonies') (January–March), pp. 52–73 (65–67), https://Al-Ādāb.com/sites/default/files/aladab_1992_v40_11_0052_0073.pdf.

Al-Amir, Daizy. 1989/1994. "Author's Preface." In *The Waiting List: An Iraqi Woman's Tales of Alienation* by Daizy Al-Amir, tr. Barbara Parmenter, ix–xii. Austin: Texas University Press in Austin.

Al-Dulaimi, Lutfiya. 1974. البشارة (قصص) [*Glad Tidings – Short Stories*]. Baghdad: Wizārat al-Thaqāfa Wa al-I'lām

Al-Dulaimi, Lutfiya. 2003/2004. حديقة حياة (رواية) [*Hayat's Garden – A Novel*]. Baghdad/Damascus. Dār al-Shu'ūn Al-Thaqāfiyya al-'Āmma/Manshūrāt Ittiḥād al-Kutāb al-'Arab.

Al-Dulaimi, Lutfiya. 2016. هاجس الهاوية في الرواية النسائية العراقية "[The Margin of Identity in the Iraqi Nisā'iyya Novel]." *Al-Arab*, 30th June 2016 [Online]. www.alarab.co.uk/?id=84030 [Accessed 10th April 2020].

Al-Halaby, Ali Hassan Ali, ed. 1999. موسوعة الحديث والآثار الضعيفة الموضوعة [*Encyclopedia of Hadith and Weakly Situated Traces*]. Riyadh: Library of Knowledge for Publishing and Distribution.

Ali, Salah Salim. 2008. "Ideology, Censorship, and Literature: Iraq as a Case Study." *Primerjalna Knjizevnost* no. 31: 213–220.

Al-Jabbar, Faleh. 1992. "Why the Uprisings Failed." *Middle East Report (MERIP)* 176, no. 1 (*Iraq in the Aftermath*): 2–14.

Al-Mana, Samira. 1997/1968. القامعون [*The Oppressors*]. Damascus: Dār Al-Madā.

Al-Mana, Samira, & Ruth Abou Rached. 2017. Informal interview, London, 17th April 2017.

Al-Marashi, Ibrahim, & Aysegul Keskin. 2008. "Reconciliation Dilemmas in Post-Ba'athist Iraq: Truth Commissions, Media and Ethno-Sectarian Conflicts." *Mediterranean Politics* 13, no. 2: 243–259.

Al-Musawi, Muhsin. 2006. *Reading Iraq: Culture and Power In Conflict*. London: I.B Tauris.

Al-Nasiri, Bouthayna. 1974. حدوة حصان [*Horseshoe*] (Short Stories). Baghdad: Dār al-Ḥurriyya.

Al-Nasiri, Bouthayna. 2000. "The Return of the Prisoner" (Short Story), tr. Denys Johnson-Davies. In *Under The Naked Sky: Short Stories from the Arab World*, selected and tr. Denys Johnson-Davies, 211–218. Cairo: American University Cairo Press.

Al-Nasiri, Bouthayna. 2001. *Final Night* (Short Stories), tr. Denys Johnson-Davies. Cairo: American University of Cairo.

Al-Qazwini, Iqbal. 2006. مرات السكون [*The Corridors of Silence*]. Amman: Dār al-Azmina Li al-Nashr Wa al-Tawzīʻ.

Al-Qazwini, Iqbal. 2008. *Zubaida's Window: A Novel of Iraqi Exile*, tr. Azza El-Kholy & Amira Nowaira. New York, Feminist Press.

Al-Radi, Nuha. 2003. *Baghdad Diaries: A Woman's Chronicle of War and Exile*. New York: Vintage.

Al-Rawi, Shahad. 2016. ساعة بغداد [*The Baghdad Clock*]. London: Dār al-Ḥikma.

Al-Rawi, Shahad. 2018. *The Baghdad Clock*, tr. Luke Leafgren. London: Oneworld Publications.

Al-Saffar, Amer. 2018. الرواية النسوية العراقية [*The Iraqi Nisūwī Novel*]. (Lecture by Najam Abdullah Kadhim in Wales, 3rd April 2018). *Aqlam/Arabic Literature*. www.aqlam.co.uk [Accessed 10th April 2020].

Alsagaff, Hussain. 2018. "The Iraqi Novel Emerges from the Womb of Disaster," tr. Jonathon Wright. In *Banipal Magazine of Modern Arab Literature 61: A Journey in Iraqi Fiction*, Vol. 1, ed. Samuel Shimon, 22–28. London: Banipal.

Al-Samman, Hanadi. 2010. قلق الاندثار: الكاتبة العربية بين ذكرى شهرازاد وكابوس المؤودة [The Anxiety of Obliteration: Arab Women Writers between the Memory of Scheherazade and the Nightmare of *Al-Mū'wuda'*]. *Alif: Journal of Comparative Poetics* 30: 73–97.

Alsanea, Rajaa. 2005. بنات الرياض [*Girls of Riyadh*]. Beirut/London: Dār Al Sāqī.

Alsanea, Rajaa. 2007. *Girls of Riyadh*, tr. Rajaa Alsanea & Marilyn Booth. New York: Penguin.

Al-Shabib, Izzadin. 2014. "Interview with Hadiya Hussein." برنامج فكر الحوار [*Dialogue of Thought Programme*], F.M.A, 11th June 2014 [Online]. www.youtube.com/watch?v=_6ZZxCnIKxQ [Accessed 10th April 2020].

Altoma, Salih J. 2010. *Iraq's Modern Arabic Literature: A Guide to English Translations Since 1950*. Lanham: Scarecrow Press.

Álvarez, Sonia E. 2014. "Enacting a Translocal Feminist Politics." In Translocalities/Translocalides, eds. Sonia E. Álvarez et al., 1–18. Durham: Duke University Press.

Al-Zayyat, Latifa. 1995. مراجعات في الأدب – حبات النفتالين ـ د. لطيفة الزيات [References in Literature – Mothballs – Dr. Latifa Al-Zayyat]. *Cairo Noor Quarterly Journal* Spring: 1–3.

Antoon, Sinan. 2010. "Bending History." *Middle East Report* 257: 29–31.

Anzaldúa, Gloria. 1987. *Borderlands: la frontera.* San Francisco: Aunt Lute.

Aref, Mohammed. 2014. يوم نساء العرب الاجنبيات "[The Foreign Arab Women's Day]." *Al-Ittihad*, 6th March 2014. www.alittihad.ae/wajhatdetails.php?id=7782 [Accessed 10th April 2020].

Arendt, Hannah. 1963/2005. *Eichmann and the Holocaust* (originally published as *Eichmann in Jerusalem, A Study in the Banality of Evil*). London: Penguin Books.

Arrojo, Rosemary. 1994. "Fidelity and the Gendered Translation." *TTR: traduction, terminologie, rédaction* 7, no. 2: 147–163.

Atia, Nadia. 2019. "Death and Mourning in Contemporary Iraqi Texts." *Interventions: International Journal of Postcolonial Studies* 21, no. 3: 1–19.

Aziz, Maher. 2011. *Kurds of Iraq, the Ethnonationalism and National Identity in Iraqi Kurdistan.* London: I.B. Tauris.

Badr, Ali. 2013. "Iraq: A Long Phantasmagorical Dream for Those Who Are Not Part of the New Capitalism or Retired Communism." In *We Are Iraqis: Aesthetics and Politics in a Time of War*, eds. Nadje S. Al-Ali & Deborah Al-Najjar, 104–118. New York: Syracuse University Press.

Bahrani, Yasmine. 2001. "'Sky' Translates an East-West Struggle." (Book Review). *USA Today*. www.betoolkhedairi.com/aboutme.htm [Accessed 10th April 2020].

Bano, Shamanez. 2015. "Women Writers from the Arab World: Fatima Mernissi, Daisy Al-Amir, Najde Al-Ali and Zainab Salbi." In *Women and Literature*: Allahabad: Allahabad University. http://epgp.inflibnet.ac.in/epgpdata/uploads/epgp_content/women_studies/gender_studies/02.women_and_literature/25._women_writers_from_the_arab_world/et/8032_et_et_25.pdf [Accessed 10th April 2020].

Baudrillard, Jean. 1991. *The Gulf War Did Not Take Place*, tr. Paul Patton. Bloomington and Indianapolis: Indiana University Press.

Beckett, Claire, & Nuit Banai. 2012. "Simulating Iraq: Cultural Mediation and the Effects of the Real." *Public Culture* 24, no. 1: 249–267.

Bhabha, Homi K. 1992. "The World and the Home." *Social Text* 31–32, no. 1: 141–153.

Bhabha, Homi K. 1994. *The Location of Culture.* London: Routledge.

Bilal, Wafaa, & Kari Lydersen. 2013. *Shoot an Iraqi: Art, Life and Resistance Under the Gun.* San Francisco City: Lights Books.

Blasim, Hassan. 2016. "Introduction." In *Iraq + 100: Stories from a Century after the Invasion*, ed. Hassan Blasim, v–x. Manchester: Comma Press.

Booth, Marilyn. 2007. "Translator v. Author (2007) Girls of Riyadh go to New York." *Translation Studies* 1, no. 2: 197–211.

Booth, Marilyn. 2008. "Translator's Note and Acknowledgements." In *The Loved Ones*, by Alia Mamdouh, tr. Marilyn Booth, 277–279. New York: Feminist Press.

Booth, Marilyn. 2010. "'The Muslim Woman' as Celebrity Author and the Politics of Translating Arabic: *Girls of Riyadh* Go on the Road." *Journal of Middle East Women's Studies* 6, no. 1: 149–182.

Brossard, Nicole, & France Theoret. 1976. "Préface." In *La nef des sorcières*, 49–50. Montreal: Quinze.

Bush, George W. 2001. "George W. Bush's Address at the Joint Session of Congress and the American People, September 21st 2001." *The Guardian*, 21st September 2001. www.theguardian.com/world/2001/sep/21/september11.usa13 [Accessed 10th April 2020].

Cahill, Mike. 2016. "They Call This the Highway of Death and You Will Cringe When You Know Why." *ViralNova*, May 11th 2016. www.viralnova.com/highway-of-death/ [Accessed 10th April 2020].

Carr, David. 2003. "A Nation at War: Bringing Combat Home *Telling War's Deadly Story At Just* Enough Distance." *New York Times*, 7th April 2003 [Online]. www.nytimes.com/2003/04/07/business/nation-war-bringing-combat-home-telling-war-s-deadly-story-just-enough-distance.html [Accessed 10th April 2020].

Castro, Olga, 2009. "Examining Horizons in Feminist Translation Studies: Towards a Third Wave?" Tr. Mark Andrews. *MonTi* 1: 1–17.

Castro, Olga, & Emek Ergun, eds. 2017. *Feminist Translation Studies: Local and Transnational Perspectives:* London/New York: Routledge.

Castro, Olga, & Emek Ergun. 2018. "Translation and Feminism." In *The Routledge Encyclopedia of Translation and Politics*, eds. Jon Evans & Fruela Fernandez, 125–143. London/New York: Routledge.

Chediac, Joyce. 2016. "25 Years Ago, 1991 Iraq Gulf War: The Massacre at the 'Highways of Death'." *Global Research: Centre for Research on Globalisation*, 27th February 2016. www.globalresearch.ca/twenty-five-years-ago-the-1991-iraq-gulf-war-america-bombs-the-highway-of-death/551840 [Accessed 10th April 2020].

Chollet, Mona. 2002. "An Interview with Alia Mamdouh." *The Handstand*, August 2002. www.thehandstand.org/archive/august2002/articles/alia.htm [Accessed 10th April 2020].

Cixous, Hélène. 2005. "Introduction" tr. Judith Miller. In *Naphtalene: A Novel of Baghdad*, by Alia Mamdouh, tr. Peter Theroux, v–vii. New York: Feminist Press.

Cockburn, Patrick. 1995. "Campaign of Mutilation Terrorises Iraqis." *The Independent*, 13th January 1995. www.independent.co.uk/news/world/campaign-of-mutilation-terrorises-iraqis-1567789.html [Accessed 10th April 2020].

cooke, miriam. 1997. "Flames of Fire in Qadisiya." In *Women and the War Story* by miriam cooke, 220–266. Berkeley: University of California Press.

cooke, miriam. 2007. "Baghdad Burning: Women Write War in Iraq." *World Literature Today* 81, no. 6: 23–26.

cooke, miriam. 2012. "Preface." In *Beyond Love*, by Hadiya Hussein, tr. Ikram Masmoudi, ix–xi. New York: Syracuse Press.

cooke, miriam, & Roshni Rustomji-Kerns, eds. 1994. *Blood into Ink: South Asian and Middle Eastern Women Write War*. Boulder/Oxford: Westview Press.

Costa, Claudia de Lima. 2014. "Feminist Theories, Transnational Translations and Cultural Mediations." In *Translocalities/Translocalidades*, eds. Sonia E. Alvarez et al., 133–148. Durham: Duke University Press.

Crenshaw, Kimberlé. 1991. "Mapping the Margins: Identity Politics, Intersectionality, and Violence Against Women." *Stanford Law Review* 43, no. 6: 1241–1299.

Dällenbach, Lucien. 1989. *The Mirror in The Text*. Chicago: Chicago University Press.

Davis, Eric. 2005. *Memories of State: Politics, History, and Collective Identity in Modern Iraq*. Berkley/London: University of California Press.

Deeb, Lara. 2005. "Living Ashura in Lebanon: Mourning Transformed to Sacrifice." *Comparative Studies of South Asia, Africa and the Middle East* 25, no. 1: 122–137.

DeGhett, Torie. 2014. "The War Photo No One Would Publish." *The Atlantic*, 8th August 2014. www.theatlantic.com/international/archive/2014/08/the-war-photo-no-one-would publish/375762/ [Accessed 10th April 2020].

De Lotbinière-Harwood, Susanne. 1991. *Re-belle et infidèle, La Traduction comme pratique de réécriture au féminin* [*The Body Bilingual: Translation as a Rewriting of the Feminine*]. Quebec: Les éditions du remue-ménage: The Women's Press.

Denike, Margaret. 2008. "The Human Rights of Others: Sovereignty, Legitimacy and 'Just Causes' for the 'War on Terror'." *Hypatia* 23, no. 2: 95–121.

Ettinger, Bracha L. 1992. "Matrix and Metramorphosis." *Differences: A Journal of Feminist Cultural Studies* 4, no. 3: 176–209.

Faqir, Fadia. 1995. "Preface." In *Mothballs*, by Alia Mamdouh, tr. Peter Theroux, v–ix. Arab Women Writers Series. Reading: Garnet Press.

Faust, Aaron M. 2015. *The Ba'thification of Iraq: Saddam Hussein's Totalitarianism*. Austin: University of Texas Press in Austin.

Flotow, Luise von, ed. 2011. *Translating women*. Ottawa: University of Ottawa Press.

Flotow, Luise von. 1991. "Feminist Translation: Contexts, Practices and Theories." *TTR: traduction, terminologie, rédaction* 4, no. 2: 69–84.

Flotow, Luise von. 2004. "Sacrificing Sense to Sound: Mimetic Translation and Feminist Writing." *Translation and Culture* 47, no. 1: 91–107.

Flotow, Luise von. 2009. "Contested Gender in Translation: Intersectionality and Metramorphics." *Palimpsestes. Revue de traduction* 22: 245–256.

Flotow, Luise von, & Farzaneh Farahzad, eds. 2016. *Translating Women: Different Voices and New Horizons*. New York/London: Routledge.

Flotow, Luise von, & Hala Kamal. 2020. *The Routledge Handbook of Translation, Feminism and Gender*, eds. Luise von Flotow & Hala Kamal. New York/London: Routledge.

Flotow, Luise von, & Carolyn Shread. 2014. "Metramorphosis in Translation: Refiguring the Intimacy of Translation Beyond the Metaphysics of Loss." *Signs: Journal of Women in Culture and Society* 39, no. 3: 592–596.

Freud Sigmund. 1919/1964. "The 'Uncanny'." In *The Standard Edition of the Complete Psychological Works of Sigmund Freud, Volume XVII* (1917–1919), ed. James Strachy, 217–256. London: Hogarth.

Freud, Sigmund. 1923/1964. "The Ego and the Id." In *The Standard Edition of the Complete Psychological Works of Sigmund Freud*, Vol. 19, ed. James Strachey, 1–66. London: Hogarth.

Frías, José Yuste. 2012. "Paratextual Elements in Translation: Paratranslating Titles in Children's Literature." In *Translation Peripheries: Paratextual Elements in Translation*, eds. Anna Gil-Bajardí, Pilar Orero, & Sara Rovira-Esteva, 117–134. Bern: Peter Lang.

Gallagher, Nancy. 1995. "Feminine and Feminist: A Review of *The Waiting List*." *Digest of Middle East Studies* 1: 63.

Gallaire, Fatima. 1990. *Les Co-épouses*. Paris: Éditions des quatre-vents.

Garrido Vilariño, Xoán M. 2005. "Texto e Paratexto: Tradución e Paratradución." *Viceversa: revista galega de traducción* 9: 31–39.

Gaul, Anny. 2014. "Translation as Mourning, Translation as a 'Form of Cultural Interrogation'." *Arabic Literature (in English)*, ed. Marcia Lynx-Qualey, 29th June 2014. https://arablit.org/2014/06/29/18571/ [Accessed 10th April 2020].

Genette, Gérard. 1987. *Seuils*. Paris: Éditions Seuil.

Genette, Gérard. 1997. *Paratexts: Thresholds of Interpretation*. Cambridge: Cambridge University Press.

Ghazoul, Ferial J. 2004. "Iraqi Short Fiction: The Unhomely at Home and Abroad." *Journal of Arabic Literature* 35, no. 1: 1–24.

Ghazoul, Ferial J. 2008. "Iraq." In *Arab Women Writers: A Critical Reference Guide, 1873–1999*, eds. Radwa Ashour, Ferial Ghazoul, & Hasna Reda-Mekdashi, 173–203. Oxford: Oxford University Press.

Gillingham, Susan. 2013. "The Reception of Psalm 137 in Jewish and Christian Traditions." *Jewish and Christian Approaches to the Psalms: Conflict and Convergence*, ed. Susan Gillingham, 64–82. Oxford: Oxford University Press.

Godard, Barbara. 1989. "Theorising Feminist Discourse/Translation." *Tessera* 6, no. 1: 42–53.

Godard, Barbara. 2002. "La traduction comme réception: les écrivaines québécoises au Canada anglais." *TTR: traduction, terminologie, rédaction* 15, no. 1: 65–101.

Hadi, Maysalun. 1994. رجل خلف الباب: مجموعة قصصية " "زينب على أرض الواقع" in *[A Man Behind the Door – Short Stories]*. Baghdad: Dār al-Shu'ūn al-Thaqāfiyya al-'Amma.

Hadi, Maysalun. 2008. "Her Realm of the Real" (Short Story), tr. Shakir Mustafa. In *Contemporary Iraqi Fiction: An Anthology*, ed. Shakir Mustafa, 68–73. New York: Syracuse University Press.

Hamedawi, Shayma. 2017. "The Postcolonial Iraqi Novel: Themes and Sources of Inspiration." *La Méditerranée au pluriel. Cultures, identités, appartenances*, 36: 211–226.

Hatto, Majeda. 2013. الرواية النسوية العراقية المعاصرة والخطاب المغاير للانهمام بالجسد وتشظي الهوية [The Contemporary Iraqi Nisūwī Novel and the Divergent Discourse on the Pre-occupation of The Body and the Fragmentation of Identity]." Baghdad, 1–24. nazikprize.crd.gov.iq/pdf/Monetary/1/11.pdf [Accessed 10th April 2020].

Hills Collins, Patricia. 2017. "Preface: On Translation and Intellectual Activism." In *Feminist Translation Studies: Local and Transnational Perspectives*, eds. Olga Castro & Emek Ergun, xi–xvi. New York/London: Routledge.

Huhn, Rosi. 1993. "Moving Omissions and Hollow Spots in the Field of Vision." In *Bracha Ettinger: Matrix-Borderlines*. Oxford: Museum of Modern Art: 5–10.

Human Rights Watch. 1992. "Endless Torment: The 1991 Uprising in Iraq and Its Aftermath." *Human Rights Watch*, June. www.hrw.org/report/1992/06/01/endless-torment/1991-uprising-iraq-and-its-aftermath [Accessed 10th April 2020].

Ismael, Jacqueline S. 2014. "Iraqi Women in Conditions of War and Occupation." *Arab Studies Quarterly* 36, no. 3: 260–267.

Ismael, Tareq. 2007. *The Rise and Fall of the Communist Party of Iraq.* Cambridge: Cambridge University Press.

James, Thomas. 1872. *Aesop' Fables: A New Version Chiefly From Original Sources.* Philadelphia: B. Lippencott and Co.

Jano, Emad, & Ruth Abou Rached. 2017. Email correspondence, 15th June 2017.

Johnson-Davies, Denys. 2001. "Introduction." In *Final Night* by Bouthayna Al-Nasiri, tr. Denys Johnson-Davies, 1–3. Cairo: American University of Cairo.

Joseph, Suad. 1991. "Elite Strategies for State-Building: Women, Family, Religion and State in Iraq and Lebanon." In *Women, Islam and the State,* 176–200. London: Palgrave Macmillan.

Kachachi, Inaam. 1998. لورنا، سنواتها مع جواد سليم[*Lorna, Her Years with Jawad Selim*]. Beirut: Al-Jadīd.

Kadhim, Sa'eed. 2017. التجارب في الرواية النسوية العراقية بعد عام ٢٠٠٣ [*Experimentalism in the Post-2003 Iraqi Nisūwī Novel*]. Baghdad: Dār al-Tamūz Li al-Ṭibā'a Wa al-Tawzī'.

Kashou, Hanan. 2013. *War and Exile in Contemporary Iraqi Women's Novels.* PhD Thesis. University of Ohio. Supervised by Professor Joseph Zeidan.

Khakpour, Arta, with Mohammad M. Khorrami, & Shouleh Vatanabadi, eds. 2016. *Moments of Silence: Authenticity in the Cultural Expressions of the Iran-Iraq War, 1980–1988, Authenticity.* New York: New York University Press.

Khodeir, Mohammed. 2013. رواية التغير في العراق: الرواية النسوية."[The Novel of Change in Iraq: The *Nisūwī* Novel]." *Al-Ṣabāḥ,* 8th July 2013. www.alsabaah.iq/ArticleShow.aspx?ID=49777 [Accessed 10th April 2020].

Khoury, Dina R. 2013. *Iraq in Wartime: Soldiering, Martyrdom, and Remembrance.* Cambridge: Cambridge University Press.

Kleinfelder, Karen. 2000. "Ingres as a Blasted Allegory." *Art History* 23, no. 5: 800–817.

Knox, Sonya. 2003. "A Sky So Close: A Journey Through War and Loss." *The Daily Star, Lebanon Culture,* 27th March 2003. www.dailystar.com.lb/Culture/Art/2003/Mar-27/111443-a-sky-so-close-a-journey-through-war-and-loss.ashx [Accessed 10th April 2020].

Kolias, Helen Dendrinou. 1990. "Empowering the Minor: Translating Women's Autobiography." *Journal of Modern Greek Studies* 8, no. 2: 213–221.

Kruk, Remke. 1993. "Warrior Women in Arabic Popular Romance: Qannasa Bint Muzahim and Other Valiant Ladies." *Journal of Arabic Literature* 14, no. 1: 213–230.

Lacan, Jacques. 1975. *Le Séminaire de Jacques Lacan, Livre XX: Encore. 1972–73,* ed. Jacques-Alain Miller. Paris: Seuil.

Leinwand, Debbie. 2003. "Iraqi Author Underscores The Commonalities of Humanity." *USA Today,* 17th February 2003. www.betoolkhedairi.com/press_en2.htm [Accessed 10th April 2020].

Levinas, Emmanuel. 1987. *Collected Philosophical Papers,* Vol. 100, tr. Alphonso Lingis. Dordrecht:Martinus Nijhoff Publishers.

Levine, Suzanne J. 1991. *The Subversive Scribe: Translating Latin American Fiction.* Minneapolis: Graywolf.

Lewental, Gershon D. 2014. "'Saddam's Qadisiyyah': Religion and History in the Service of State Ideology in Ba'thist Iraq." *Middle Eastern Studies* 50, no. 1: 891–910.

Loucif, Sabine. 2012. "Lectures d'aujourd'hui aux USA: les dessous du marché de la traduction." In *Narrations d'un nouveau siècle. Romans et récits français (2001–2010)*, eds. Bruno Blanckeman & Barbara Havercroft. Paris: Presses Sorbonne Nouvelle.

Lynx-Qualey, Marcia. 2010. "The American Granddaughter: British vs. American Dialect in Translating 'The American Granddaughter'." *Arabic Literature (in English)*, ed. Marcia Lynx-Qualey, 29th October 2010. https://arablit.org/2010/10/29/british-vs-american-dialect-in-translating-the-american-granddaughter/ [Accessed 10th April 2020].

Lynx-Qualey, Marcia. 2013. "'I feel closer to my country when I'm away': Interview with Hadiya Hussein." *Qantara.de*, 4th April 2013. https://en.qantara.de/content/interview-with-iraqi-author-hadiya-hussein-i-feel-closer-to-my-country-when-im-away [Accessed 10th April 2020].

Lynx-Qualey, Marcia. 2014. "Inaam Kachachi on 'Tashari' and the Iraq She Carries With Her." *Arabic Literature (in English)*, ed. Marcia Lynx-Qualey, 4th February 2014 [online]. https://arablit.org/2014/02/04/inaam-kachachi-on-tashari-and-the-iraq-that-she-carries-with-her/[Accessed 10th April 2020].

Maier, Carol. 1994. "Afterword: 13 Glosses." In *Memoirs of Leticia Valle*, by Rosa Chacel, tr. Carol Maier, 165–194. Lincoln: University of Nebraska Press.

Malti-Douglas, Fedwa. 1991. *Woman's Body, Woman's Word: Gender and Discourse in Arabo-Islamic Writing*. Princeton: Princeton University Press.

Mamdouh, Alia. 1998. "Creatures of Arab Fear" tr. Shirley Eber & Fadia Faqir. In *In The House of Silence: Autobiographical Essays by Arab Women Writers*, ed. Fadia Faqir, 63–71. Reading: Garnet Publishing.

Mamdouh, Alia. 2004. "Baghdad: These Cities Are Dying In Our Arms" tr. Marilyn Booth, *Word: On Being a [Woman] Writer*, Vol. 0, ed. Jocelyn Burrell, 38–48. New York: Feminist Press.

Mamdouh, Alia. 2008. *The Loved Ones*, tr. Marilyn Booth. New York: Feminist Press.

Masmoudi, Ikram. 2010. "Portraits of Iraqi Women: Between Testimony and Fiction." *International Journal of Contemporary Iraqi Studies* 4, no. 1–2: 59–77.

Masmoudi, Ikram. 2012. "Introduction." In *Beyond Love*, by Hadiya Hussein, tr. Ikram Masmoudi, xv–xx. New York: Syracuse Press.

Masmoudi, Ikram. 2015. *War and Occupation in Iraqi Fiction*. Edinburgh: Edinburgh University Press.

Massardier-Kenney, Françoise. 1997. "Towards a Redefinition of Feminist Translation Practice." *The Translator* 3, no. 1: 55–69.

Mattar, Karim. 2019. "Interview with Sinaan Antoon." In *The Edinburgh Companion to the Postcolonial Middle East*, eds. Anna Ball & Karam Mattar, 67–80. Edinburgh: Edinburgh University Press.

Mattar, Mohamed Y. 2011. "Human Rights Legislation in the Arab world: The Case of Human Trafficking." *Michigan Journal of International Law* 33: 101–132.

McCann-Baker, Annes. 1994. "Acknowledgements." In *The Waiting List: An Iraqi Woman's Tales of Alienation* by Daizy Al-Amir, tr. Barbara Parmenter, vii. Austin: University of Texas Press in Austin.

Mehta, Brinda. 2006. "Dissidence, Creativity, and Embargo Art in Nuha Al-Radi's Baghdad Diaries." *Meridians: Feminism, Race, Transnationalism* 6, no. 2: 220–235.

Mikhail, Mona. 1994. "Introduction." *The Waiting List: An Iraqi Woman's Tales of Alienation* by Al-Amir, Daizy, tr. Barbara Parmenter, 1–5. Austin: University of Texas Press.

Minh-Ha, Trinh. T. 1995. "No Master Territories." In *The Post-Colonial Studies Reader*, eds. Bill Ashcroft Gareth Griffiths & Helen Tiffin, 215–218. London: Routledge.

Moosavi, Amir. 2015. "How to Write Death: Resignifying Martyrdom in Two Novels of the Iran-Iraq War." *Alif: Journal of Comparative Poetics* 35, no. 1: 9–31.

Mushatat, Raad. 1986. "At Home and in Exile." *Index on Censorship* 15, no. 2: 28–31.

Mustafa, Shakir, ed. 2008. *Contemporary Iraqi Fiction: An Anthology*. New York: Syracuse University Press.

Mustafa, Shakir. 2004. "Two Stories by Samira Al-Mana." *Edebiyāt: Journal of Middle Eastern Literatures* 14, no. 1–2: 129–131.

Najjar, Al-Mostapha. 2014. "Inaam Kachachi: 'We Are Experiencing a True Upsurge in Iraqi Fiction'." *Arabic Literature (in English)*, ed. Marcia Lynx-Qualey, 18th April 2014. https://arablit.org/2014/04/18/inaam-kachachi-we-are-experiencing-a-true-upsurge-in-iraqi-fiction/ [Accessed 10th April 2020].

Niazi, Salah, & Ruth Abou Rached. 2020. Informal interview, London, 25th January 2020.

Niazi, Salah, & Samira Al-Mana, eds. 1985–2002. لاغتراب الأدبي ا [*Literature in Exile*]. London: Al-Ightirab Al-Adabi.

Pedersen, Marianne. 2014. *Iraqi Women in Denmark: Ritual Performance and Belonging in Everyday Life*. Oxford: Oxford University Press.

Phelan, Peggy. 1993. *Unmarked: The Politics of Performance*. New York: Routledge.

Phillips, George. 1846. *The Psalms in Hebrew: With a Critical, Exegetical, and Philological Commentary*, Vol. 2, 552–553. https://archive.org/details/psalmsinhebrewwi02phil [Accessed 10th April 2020].

Reimóndez, María. 2017. "We Need to Talk . . . to Each Other: On Polyphony, Postcolonial Feminism and Translation." In *Feminist Translation Studies: Local and Transnational Perspectives*, eds. Olga Castro & Emek Ergun, 42–55. New York/London: Routledge.

Robert-Zibbels, Elizabeth. 2002. "Betool Khedairi's A Sky So Close." *eserver_reconstruction* http://reconstruction.eserver.org/BReviews/revSkySoClose.htm [Accessed 10th April 2020].

Rohde, Achim. 2010. *State-Society Relations in Ba'thist Iraq: Facing Dictatorship*. London: Routledge.

Sadiqi, Fatima. 2006. "Gender in Arabic." In *Brill Encyclopaedia of Linguistics*, 1–21. Brill (online).

Safouan, Moustapha. 2007. *Why Are the Arabs Not Free? The Politics of Writing.* Oxford: Blackwell.

Said, Edward. 1990. "Embargoed Literature." *The Nation* 17, no. 1: 278–280.

Salloum, Sa'ad, ed. 2013. *Minorities in Iraq: Memory, Identity and Challenges,* Translation into English by The Syrian European Documentation Centre Baghdad-Beirut, Masarat. Baghdad: National Library and Archive (Baghdad: 2041/2012).

Šešić, Milena, Sprko Leštarić, & Edward Alexander. 2014. *12 Impossibles: Stories by Rebellious Arab Writers.* Preface Milena Šešić. Selected and tr. from Arabic into Serbian Sprko Leštarić (2005). Tr. Serbian into English by Edward Alexander. Sofia: European Cultural Foundation and Next Page Foundation. www.cultural foundation.eu/library/12-impossibles [Accessed 10th April 2020].

Shabbar, Fatin. 2014. "Motherhood as a Space of Political Activism." In *Motherhood and War: International Perspectives,* ed. Dana Cooper & Claire Phelan, 207–224. New York: Palgrave Macmillan.

Shabout, Nada. 2013. "Bifurcations of Iraq's Visual Culture." In *We are Iraqis: Aesthetics and Politics in a Time of War,* eds. Nadje Al-Ali & Deborah Al-Najjar, 6–23. New York: Syracuse University Press.

Shread, Carolyn. 2007. "Metamorphosis or Metramorphosis? Towards a Feminist Ethics of Difference in Translation." *TTR: traduction, terminologie, rédaction* 20, no. 2: 213–242.

Shread, Carolyn. 2011. "On Becoming in Translation." In *Translating Women,* ed. Luise von Flotow, 283–304. Ottawa: University of Ottawa Press.

Simawe, Saadi A. 2004. "Iraq Over the Transom." (Interview by Nataša Ďurovičová, dated early June). *91st Meridien* 3, no. 3. Iowa City: University of Iowa. https://iwp.uiowa.edu/91st/vol3-num3/iraq-over-the-transom [Accessed 10th April 2020].

Simawe, Saadi A. 2009. "Contemporary Iraqi Fiction: An Anthology (Book Review)." *Journal of Arabic Literature* 40, no. 1: 129–132.

Simon, Sherry. 1996. *Gender in Translation.* New York/London: Routledge.

Snaije, Olivia. 2014. "Exiled in Europe: An Interview with Three Women Writers." *Magazine: Words Without Borders,* September 2014. www.wordswithoutborders. org/article/exiled-in-europe-an-interview-with-three-women-writers

Spivak, Gayatri C. 1988. "Can the Subaltern Speak?" In *Marxism and the Interpretation of Culture,* eds. Cary Nelson and Lawrence Grossberg, 271–315. Urbana: University of Illinois Press.

Spivak, Gayatri C. 1993. "The Politics of Translation." In *Outside in the Teaching Machine,* 179–200. London and New York: Routledge.

Spivak, Gayatri C. 2000. "A Moral Dilemma." *Theoria: A Journal of Social and Political Theory,* 96: 99–120.

Strochlic, Nina. 2017. "Famed 'Afghan Girl' Finally Gets a Home." *National Geographic,* 12th December 2017 www.nationalgeographic.com/news/2017/12/afghan-girl-home-afghanistan/ [Accessed 10th April 2020].

Talib, Aliya. 1994. "A New Wait" (Short Story), tr. miriam cooke & Rkia Cornell. In *Blood into Ink: South Asian and Middle Eastern Women Write War,* eds. miriam cooke & Roshni Rustomji-Kerns, 80–85. Boulder/Oxford: Westview Press.

Thamir, Fadil. 2004. الرواية العراقية من الريادة الى النضج "[The Iraqi Novel from Pioneering to Maturity]." *Culture. Al-Madā Al-Thaqāfī,* 18th August 2004.

Toorawa, Shawkat M. 2007. "The Medieval Waqwaq Islands and the Mascarenes." *The Western Indian Ocean: Essays on Islands and Islanders*, 1st ed., 49–66. Port Louis: The Hassam Toorawa Trust.

Tramontini, Leslie. 2013. "'Speaking Truth to Power?' Intellectuals in Iraqi Baathist Cultural Production." *Middle East – Topics and Arguments* 1: 53–61.

VanderKam, James C. 1985. "The Book of Jubilees." In *Outside the Old Testament*, ed. Marinus De Jonge, 111–144. Cambridge: Cambridge University Press.

Voykowitch, Brigitte. 2003. "Book Review: A Sky So Close." *Die Gazette*, 10th May 2003, https://gazette.de/Archiv/Gazette-Mai2003/Khedairi.html [Accessed 10th April 2020].

Wali, Najem. 2007. "Iraq" tr. Lilian M. Freidberg. In *Literature from the 'Axis of Evil': Writing From Iran, Iraq, North Korea, and Other Enemy Nations*, 51–54. Words Without Borders /The New Press.

Wallmach, Kim. 2006. "Feminist Translation Strategies: Different or Derived?" *Journal of Literary Studies* 22, no. 1–2: 1–26.

Wehr, Hans. 1994. *A Dictionary of Modern Written Arabic*, ed. J.M. Cowan. New York: Ithaca.

Winegar, Jessica. 2005. "Of Chadors and Purple Fingers: US Visual Media Coverage of the 2005 Iraqi Elections." *Feminist Media Studies* 5, no. 3: 391–395.

Wing, Joel. 2013. "Iraqi Women Before and After The 2003 Invasion, Interview with Prof. Nadje Al-Ali University of London." *Musings on Iraq: Iraqi News, Politics, Economics, Society*, 23rd December 2013. http://musingsoniraq.blogspot.co.uk/2013/12/iraqi-women-before-and-after-2003.html [Accessed 10th April 2020].

Youssef, Faruq. 2010. "الحفيدة الأميركية : كجه جي تضع يدها على جمرة الهلاك العراقي" [Inaam Kachachi Puts Her Hand on the Burning Ember of Iraq's Destruction]. *Nuqtat al-daw' al-yawm* [*In the Spotlight Today*], 10th April 2020. www.masress.com/ndawa/290 [Accessed 10th April 2020].

Zangana, Haifa. 1990. *Through the Vast Halls of Memory*, tr. Paul Hammond & Haifa Zangana. France and Basingstoke: Hourglass Press.

Zangana, Haifa. 1995. في اروقة الذاكرة [*In the Corridors of Memory*]. London: Dār Al-Hikma.

Zangana, Haifa. 2004. "Introduction." In *Women on a Journey: Between Baghdad and London*, by Haifa Zangana, tr. Judy Cumberbatch, ix–xvi. Austin: University of Texas in Austin.

Zangana, Haifa. 2005. "Colonial Feminists from Washington to Baghdad." *Al-Raida Journal* 22, no. 109–110: 30–40.

Zangana, Haifa. 2009. *Dreaming of Baghdad*, tr. Haifa Zangana & Paul Hammond, 169. New York: Feminist Press.

Zangana, Haifa. 2014. هل هناك مبدعات عراقيات حقا؟ "[Are There Really Any Iraqi Women Creatives Out There?]." *Al-Quds Al-'Arabi*, 22nd April 2014. www.alquds.co.uk/?p=159251 [Accessed 10th April 2020].

Zangana, Haifa, & Ruth Abou Rached. 2018. Email correspondence, 15th March 2018.

Zimbardo, Philip G. 2007. *Lucifer Effect*. Oxford: Blackwell Publishing Ltd.

Anonymous sources

"Si je t'oublie, Bagdad." *FranceCulture*. www.franceculture.fr/oeuvre-si-je-t-oublie-bagdad-de-inaam-kachachi.html [Accessed 10th April 2020].

"Si je t'oublie, Bagdad d'Inaam Kachachi." http://missbouquinaix.com/2013/05/06/si-je-toublie-bagdad-dinaam-kachachi-2008 [Accessed 10th April 2020].

Index

Printed in the United States
by Baker & Taylor Publisher Services

Printed in the United States
by Baker & Taylor Publisher Services